TOMATOES

Margaret Gin
Drawings by Rik Olson

101 Productions
San Francisco

Copyright © 1977 Margaret Gin
Drawings copyright © 1977 Rik Olson

All rights reserved. No part of this book may be reproduced in any form without the permission of 101 Productions.

Printed and bound in the United States of America.

Distributed to the book trade in the United States by Charles Scribner's Sons, New York, and in Canada by Van Nostrand Reinhold Ltd., Toronto

Published by 101 Productions
834 Mission Street
San Francisco, California 94103

Library of Congress Cataloging in Publication Data

Gin, Margaret.
   Tomatoes.

   (Edible garden series)
   Includes index.
   1. Cookery (Tomatoes) 2. Tomatoes. I. Title.
II. Series.
TX803.T6G56   641.6'5'642   77-3481
ISBN 0-89286-111-8 pbk.

contents

Tomato Talk 4
The Tomato Patch 8
Soups & Drinks 24
Salads 32
Hot Side Dishes 44
Main Dishes 52
Sauces 68
Sweets & Pastries 76
Preserving 80
Index to Recipes 94

tomato talk

Called "xtomatle" by the Aztecs, the tomato *(Lycopersicon esculentum)* was discovered during the 16th century by the Spanish conquistadors in their search for Inca treasures. They called them *manzanas* meaning "golden apple." These early tomatoes were small and yellow and though decorative were thought to be poisonous because they were related to the deadly nightshade. The Spaniards were fascinated by their beauty and brought them back with them to Europe as ornamental plants. The tomato was not used as food until a century later. In a relatively short period of time it has become such an essential ingredient in so many cuisines that it is hard to imagine cooking without it.

The Spanish Moors introduced the tomato to Italy and the Italians called it *pomo del moro* or "Moor's apple." Pomo del moro found its way to France, where the Italian name sounded so much like *pomme d'amour* that the Moor's apple became a love apple. The French believed their *pomme d'amour* had aphrodisiac powers, and in early times the tomato was given as a love token.

The tomato has been a common garden plant in the United States since Colonial days. The home-grown, vine-ripened tomato is so superior to today's green-picked, tasteless market tomato that it is the favorite plant of the home gardener. The memory of the taste of "real" tomatoes, and the difficulty of finding them, has tempted many people into their first experiments with gardening. Tomatoes are easy to grow and can produce abundantly for very little effort and expense. As few as six to 12 plants can supply a family of four with a year's supply of tomatoes.

The taste of a home-grown tomato seems to be the essence of summer, but freezing, canning and drying will allow you to feast on tomatoes year round. They can be used in a multitude of dishes, including soups, salads, sauces, vegetable dishes, entrées and even desserts, such as green tomato pie. Because of their mild acidity and high nutrient content, tomatoes are even used as an ingredient in skin-care products, such as soaps, creams and lotions.

Tomatoes make happy marriages with many other ingredients. Some of these are: veal, chicken, seafood, beef, lamb, pork (including bacon, sausages and ham), eggs, anchovies, cheese, lemon, onions and garlic, olive oil, butter, yoghurt, sour cream, mayonnaise and vinegar. All of these herbs and spices are delicious with tomatoes: basil, bay leaf, celery seed, chervil, chili powder, cinnamon, chives, coriander, curry, cumin, dill, ginger, marjoram, mint, mustard, oregano, parsley, rosemary, sage, tarragon and thyme.

High in vitamin content and low in calories, the tomato is one of the most nutritious of foods. One medium-sized tomato provides 57 percent of the recommended daily allowance of vitamin C for an adult, more than one-fourth of the vitamin A allowance and 1/12 of the iron, yet contains only 35 calories. Tomatoes retain vitamin C well and gain in this vitamin as they ripen. The more direct sun a tomato gets during its growth period, the greater the amount of vitamin C it will contain.

PEELING TOMATOES

To peel or skin tomatoes, begin with a pot of briskly boiling water. Add only a few tomatoes at a time and roll them around in the constantly boiling water for about 30 seconds. Remove the tomatoes from the water and dip in cold water immediately to prevent further cooking. (Some cooks prefer dipping them in ice water, especially for raw tomato dishes.) The skins will slip off easily. Cut off stem end and core. An alternate method of peeling is to pierce a tomato with a fork and hold it over a direct flame, such as on a gas stove, until the skin cracks. Remove immediately from over heat and peel off skin. This method is good for one or two tomatoes. However, if you have many tomatoes to peel, I do not recommend it. In general, when preparing tomatoes for use raw it is necessary to peel the skin only if it is tough or thick. Thin-skinned tomatoes can be used with the skin left on, which often lends additional flavor.

SEEDING TOMATOES

Halve or quarter tomatoes. With the half or quarter in your palm, hold it upside down over the sink or a container, gently squeezing and shaking the tomato until the seeds have fallen out. You may have to help a bit by scooping out the seed pockets with a small spoon.

PREPARING TOMATOES FOR STUFFING

Select large firm, ripe tomatoes for stuffing. Cut off the top half-inch of the tomato and scoop out the pulp and seeds to make a shell. Invert on a colander for 20 to 30 minutes to drain off excess moisture. (If the tomatoes are to be stuffed and baked, lightly salt the inside of the shell before inverting.) If a recipe does not call for the removed pulp, use it in sauces, soups, stews, etc.

STORING TOMATOES
Store ripe tomatoes, stem end down, in a warm place (about 60°F is best). Keep them out of sunlight. For storing end-of-the-season tomatoes while they are ripening, see Harvesting.

EQUIVALENTS
1 pound tomatoes equals 2 large, 3 medium or 4 small tomatoes
1 to 1-1/4 pounds tomatoes equals 2 cups peeled and chopped tomatoes
2-1/2 pounds tomatoes equals 1 quart peeled and chopped tomatoes
25 pounds (about 2 pecks) raw tomatoes yields about 10 quarts canned tomatoes
50 pounds (about 1 bushel) raw tomatoes yields about 20 quarts canned tomatoes

USING THIS BOOK
In the recipes that follow all tomatoes are to be standard-size, red ripe, firm ones unless otherwise specified.

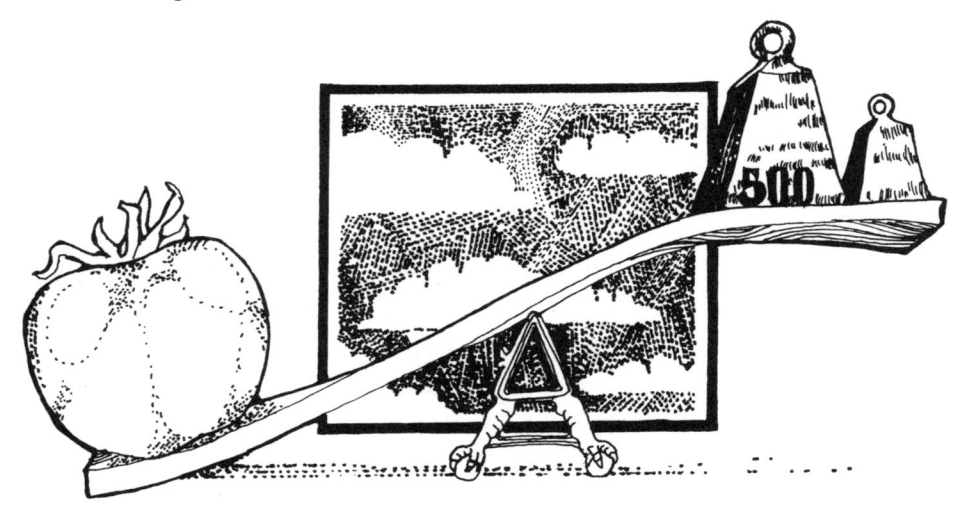

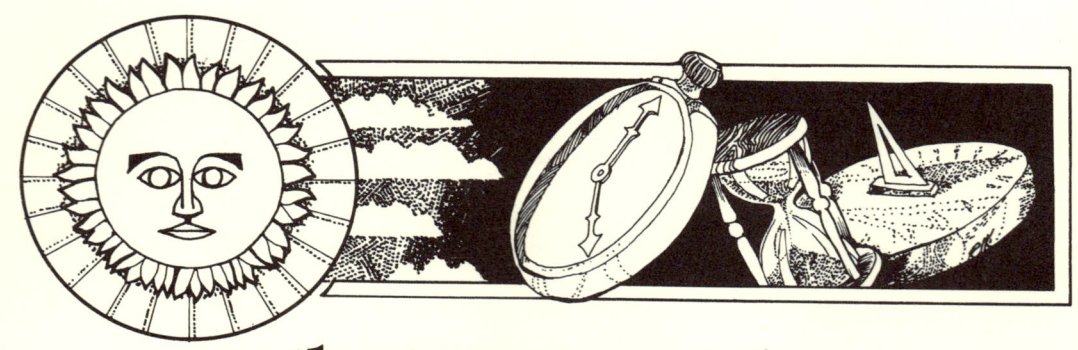

the tomato patch

VARIETIES

When choosing a tomato variety to grow in your garden, consider your needs and preferences: Do you want to harvest tomatoes for a short or long period (early or determinate type or late or indeterminate type)? How much space can you devote to your tomato plants? Are your food needs better served by an all-purpose eating and cooking tomato or one for canning, sauces, catsup, pickling, etc.? These are just some of the possible questions.

Tomato varieties belong to one of two basic classifications: determinate and indeterminate. The distinction between these is based on growth patterns. The determinates are the early varieties. They are the bush or sprawling types with the fruit generally borne low on the plant. Determinates should not be pruned as pruning cuts the yield of the plant, and they seldom need staking. All of the fruit sets at about the same time, and once it does the plant stops growing. Harvest time is short. The indeterminates respond to both pruning and staking. They are the late- or main-crop types and continue to grow and bear fruit over a long period of time. The terminal bud of an indeterminate produces leaves and more stem, unlike the determinate. These vines will grow indefinitely if not hit by the cold. Flower clusters and fruit form progressively so you will have both blossoms and ripe fruit on the same vine at the same time.

Tomato varieties have been developed which are best adapted for the various climate zones. Your neighbors, local garden center or farm bureau should be able to offer information on the availability of these varieties in your area. There are also the "All-American" varieties which have been USDA tested and pronounced suitable for growing in all parts of the United States. Most of the new hybrids are disease-resistant and are also excellent choices (see Tomato Troubles). If a short growing season is a concern, consider a quick-maturing variety; depending on variety, tomatoes require about 60 to 85 days to reach maturity from seedlings.

Taste will be a factor in your choice also. Standard, all-purpose tomatoes such as Beefsteak, Early Salad, Better Boy, Big Boy, Moreton Hybrid, Oxheart and Wonder Boy are favorites. Mature tomatoes come in an array of colors, from white (Snow White) to green (Evergreen). There are also yellow, orange and blue ones, though red will remain the preferred color, no doubt, forever. Generally speaking, the lighter-colored varieties are milder in taste.

For thick tomato sauces, purées, paste and catsup, the plum tomato varieties are the best choice for their meatier pulp and good flavor with fewer seeds. Cherry tomatoes take a minimum of space to grow and are excellent choices for containers or for growing indoors on a windowsill.

You may also decide to grow relatives of the common tomato that belong to other species. The subtropical tree tomato does best in warm climates where it will bear for 6 months a year, grow to a height of 8 feet and yield 40 to 60 pounds of fruit a season. The small tart fruit that is produced is a favorite for preserves and the tree will continue to produce year after year. (There is also a domestic tree tomato which produces very large fruit on an oversized indeterminate vine.) The most common tomatoes with lightweight husks are the husk tomato and the tomatillo. The former resembles in appearance the Chinese lantern plant and the yellow-green fruit is covered with a paper-like husk that turns brown at harvest time. Husk tomatoes are eaten raw or cooked, often as preserves, and store well. The husk tomato is also commonly known as the ground cherry, poha berry, strawberry tomato and dwarf cape gooseberry. The tomatillo is a close relative of the husk tomato and is particularly popular in Latin American dishes and sauces, both raw and cooked. It grows in the same climatic and soil conditions as regular tomatoes and is self-seeding. The small, round green fruit requires neither peeling nor seeding, and is at its best just as the husk turns brown. It can, however, be left on the vine; its tartness will lessen and its skin will turn golden. Tomatillos can be stored in the husk for several months, either on the vine or removed.

STARTING SEEDS INDOORS

Start tomato seeds 8 to 10 weeks before the date of the last frost in your area. A variety of containers may be used: wooden or plastic flats, aluminum loaf pans, cut-down milk cartons, cut-down bleach bottles, flower pots, peat pots, etc. The basic requirements are that the container have drainage holes (except peat pots), added by you if necessary, and that it be at least 2-1/2 inches deep to allow for ample root growth.

Prepare the containers for the seed by putting a thin layer of leaf mold in the bottom. Lightly dust this layer with bonemeal or well-rotted manure to encourage deep, vigorous root growth. Then fill the container to 1/2 inch from the rim with one of the following: a lightweight, fast-draining commercial soil mix designed for seed starting; a mixture of half peat moss and half vermiculite; your own soil mix made by combining equal parts builders' sand, sterilized loam and well-decayed compost or leaf mold and sifting it through 1/4-inch mesh screening. Water the growing medium well and smooth the surface. Broadcast the seeds or place them in drills over the surface so that they are 1 inch apart. Cover the seeds with a 1/4-inch thick layer of vermiculite; this will help to retain moisture in the soil and cut down the frequency of watering during germination. Lightly water top layer, being careful not to disturb the seeds. Cover the container with newspaper or plastic and keep in a warm place until the first sign of growth pokes through the soil; this should take about a week.

The optimum temperature for germination is about 70°F. To maintain this level, heating cables can be placed beneath the soil in the flat. These are easy to use and available at nurseries and garden centers, with directions for their use. They provide even heat which will speed germination and ensure maximum plant growth. A waterproof household heating pad set at medium and placed under the container, or a food warming tray set at low can also be used. Check the containers each day to make sure they have not dried out or that the condensation buildup is not too great. If there is too much moisture, mold may grow; allow air to circulate to evaporate some of the condensation.

When the seedlings are up, remove the newspaper or plastic and keep in full sunlight for as many hours a day as possible; a sunny window is a good location. Optimum temperatures at this stage are 70° to 75°F during the day and 60° to 65°F at night. Turn the containers every day to prevent the seedlings from growing only in the direction of the sun. Always keep the soil moist; it must never dry out during this critical growth period.

If you cannot provide adequate sunlight, you can use special plant-growth lights and supply 14 to 16 hours of light each day. With this overhead illumination, turning the containers will not be necessary and you will have control over the amount of light the plants receive. It will be necessary, however, to be more diligent about watering, as the soil dries out faster.

When the second pair of leaves appears, it is time to give the plants more space to grow. Mix a small amount of bonemeal and well-rotted manure or other organic fertilizer with some rich loam and put it in the bottom of a container at least 3 inches deep. Select the largest, healthiest-looking seedlings for transplanting. They should be spaced far enough apart, about 2 inches, so they can be pricked out without disturbing the root systems of other prospective transplants. It is a good idea to carefully pinch off any seedlings except the ones you want to transplant. To remove the seedling, slip a dibble or other pencil-like implement into the soil at a 45-degree angle and pop the seedling out into your hand. Set the seedling in the new container and cover with additional soil mix up to the first leaf, covering part of the main stem. New roots will grow from the buried stem and will encourage a stronger main stem. This is very beneficial to the plant when planted outdoors. The seedlings should be placed at least 3 inches apart in the new containers. They should continue to receive the maximum amount of light possible and sufficient water.

When new growth appears, and there are 4 to 6 leaves, it is time to begin hardening off the seedlings. This process conditions them for growing outdoors and should be started about a week to 10 days before setting out the seedlings to their permanent locations. The first step is to reduce watering. The soil should not be as moist as before and during germination. Do not allow the leaves to wilt, but water only when the soil begins to feel slightly dry. Reducing the amount of water will inhibit growth, thus preventing plants from getting too leggy. Also, you will want to set the plants in a cooler spot at night than during the early growing stages, about 50°F. During the day, continue to give them all the sunshine possible, remembering that tomatoes are warmth-loving plants.

Start taking the plants outdoors for limited time periods. Begin the first day by exposing the plants to only 15 minutes in a sunny sheltered area. On the next day, increase it to 30 minutes, the following day 1 hour, and continue doubling the amount of time until the plants can withstand the day and night temperatures of outdoors. At this point, the plants are ready to be transplanted outdoors permanently. Be certain all danger of frost has passed.

A commonly used device for hardening off seedlings is the cold frame. This is a bottomless box, usually made of wood, with a removable glass or plastic cover. The top of the frame slants so that the front is about 2 inches lower than the rear. Wooden fruit boxes with their bottoms knocked out can be sunk in the garden soil at an angle for use as cold frames; old windows or heavy plastic sheeting can be used for covers. The tomato seedlings are placed in the cold frame, and the cover is removed or propped up by a stick for longer and longer periods each day as the plants become acclimated to the outdoor temperatures.

Seeds may also be started in containers that can hold them until they reach the size to be transplanted to permanent outdoor locations. Sow the seeds in 3- or 4-inch deep containers; use the same soil mix prescribed for the initial transplant in the first seed starting method, or a commercial starting mix, sowing seeds as directed. Cover with plastic and keep in 75°F temperatures in a bright location. Keep the soil damp but not soaked. Harden off seedlings as described above.

Some gardeners feel the ideal containers to use for this method of seed starting are peat cubes or peat pots. With the cubes, you don't even need to add soil; the seed is dropped into a hole in the top of the cube, the cube is watered and the seedling grows. When it is time to plant outdoors, the seedling is planted, cube and all. Both the peat cube and peat pot are designed to decompose in the soil, though gardeners have reported finding the peat containers intact when digging up the garden at the end of the season.

PURCHASING SEEDLINGS

You can of course dispense with starting your own seeds and instead purchase tomato seedlings at a local nursery or garden center. If you do this, it is likely that the varieties they offer will be the ones best suited to growing in your area. Try to select healthy, stocky plants with 4 to 6 sets of leaves. If you find yourself faced with rows of spindly looking plants, don't be discouraged. Tomato plants may be planted deeply, as roots will form all along their stem (see Planting Seedlings Outdoors).

WHERE TO PLANT

Whether you grow tomatoes in containers or in the ground, the location must have ample sunlight. The more sun and thus warmth, the sweeter the tomato and the greater the amount of vitamin C. Tomatoes must have at least 6 hours of sun daily, though 12 hours or more will give you a superior fruit. Tomatoes will not yield well if grown in a shady area, but can produce an adequate crop with less than optimum sunlight. Warmth is their greatest need.

Tomatoes should never be grown in soil where plants of the same family, *Solanaceae* or nightshade, have previously been grown. These include potatoes, eggplant, peppers, okra, petunias, salpiglossis, dahlias, bittersweet, jimson weed, tobacco and Japanese lanterns. The danger is that these crops may carry over diseases onto the tomato plants. Also, do not plant tomatoes near shallow-rooted trees or large shrubs, as the plants will compete for soil nutrients, or in the vicinity of walnut trees, as the roots of the walnut tree have a poisonous effect on tomato roots.

Tomatoes do best in night temperatures of 55° to 75°F and daytime temperatures between 75° and 80°F. A southern-exposed wall or construction of some kind can help maintain these temperatures. It will also protect the tomato plants from strong winds and lengthen the growing season.

PREPARING THE SOIL

Tomatoes will grow well in a variety of soils—sandy, loam, even heavy clay—as long as the soil holds moisture, has good drainage, is rich in nutrients and retains warmth. However, the best soil is light, sandy loam that is slightly acid with a pH range of 6.0 to 6.8. The best way to improve your soil is to add compost; it is an enricher and improves the texture "naturally." You can make your own compost or purchase it packaged from the garden center. An ideal homemade compost formula is 4 parts plant waste material, such as hay, leaves, plant cuttings and vegetable debris to 1 part well-rotted manure, with a sprinkling of bonemeal, egg shells, bloodmeal or wood ashes; allow this to decompose completely before application.

To grow the best tomatoes, work the ground in the fall. Dig the soil to a depth of at least 8 inches and add a 3-inch layer of compost. Cover the cultivated area with a 6-inch layer of mulching material (see Mulching) and secure mulch with chicken wire if necessary. When you sow your tomato seeds, about 8 to 10 weeks before planting out the seedlings, it is time to work the soil again. Pull off the mulch and rake in additional compost or a 5-10-10 fertilizer, about 1 pound for every 150 square feet. Replace mulch and leave it on until 2 weeks before setting out seedlings. Then remove the mulch completely and allow the soil to warm up. In 2 weeks time you will be able to plant your seedlings.

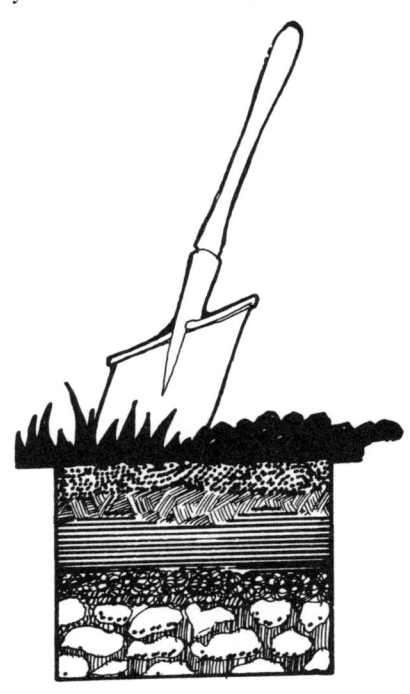

If you do not prepare your soil in the fall, as soon as the weather permits work the soil in early spring by adding compost and digging as for fall. Cover with mulch until 2 weeks before setting out seedlings and then remove mulch to allow soil to warm up.

If you follow these soil-preparation practices your soil will be rich enough in nutrients that fertilizing your tomato plants once they are established will not be necessary. If, however, you find that your plants could use a nutrient boost or you want to ensure a bumper crop, try one of the organic fertilizer mixes on the market. Be sure to select one that will complement your soil and read the directions carefully before applying it in your garden.

PLANTING SEEDLINGS OUTDOORS

The best time of day to plant tomatoes is in the evening when the sun has begun to set, or on a cloudy day. Plant bush or sprawling types 3 to 4 feet apart and the staked varieties 2 feet apart. You may plant them in a row or randomly spaced. Dig a hole twice as wide and twice as deep as the root ball. If using a simple single staking method, insert a stake on one side of the hole. (For additional staking methods, see Staking.) Place a trowelful of organic fertilizer made from 1 part phosphorus and 2 parts bloodmeal or well-rotted manure and 4 parts wood ashes into the hole. Then combine half of the soil from the hole with an equal amount of fertilizer.

Flood the hole gently with water. Set the root ball deep enough into the hole so that the plant will eventually be covered with soil up to its first set of leaves. Cover the roots with the mixture of soil and organic fertilizer, being careful not to damage the roots. Firm surface soil to remove any air pockets. Water gently and thoroughly.

If transplanting seedlings growing in peat pots, tear a strip from the top of the pot before setting it in the hole. If this precaution is not taken, the pot will act as a wick and cause the root ball to dry out. If the seedling is in a peat cube, simply place it in the hole. Fertilize the soil first as described above.

Water plants daily for the first few days after planting to prevent them from wilting. After this initial period, water only when the soil is slightly dry to the touch; it should always contain some feeling of moisture. It is also important to water deeply to benefit the roots.

Tomato plants need warm sunny day and night temperatures between 60° and 70°F to set fruit. Long spells of rainy or cloudy weather and temperatures that drop below 58°F can be hazardous to bloom and fruit may not set. A variety of constructions can be erected to protect the young tomato plants against unexpected frost, severe wind or to increase night temperatures. Fiber glass panels or sheets of plastic placed over arched wire frameworks and set over the plants will ensure protection from most climate changes. A long rectangular-shaped framework with plastic sheeting, bleach bottles with the bottoms cut out, up-ended clay pots, glass half-gallon wine bottles with the bottoms knocked out and the cork or cap discarded, or four-sided cardboard shields slipped over the plants will also all afford adequate protection.

There are also hormone fruit-setting sprays which, when sprayed on blossoms just as they begin to open, may offset adverse weather conditions. Sprayed tomatoes will ripen 1 to 3 weeks earlier than those not treated. Sprays are effective in early spring when temperatures drop below 60°F. They will not hasten ripening during periods of warm nights and 90°F or warmer days.

MULCHING

Mulching your tomato plants will prevent the soil from drying out. The kind of mulch you use will depend upon cost and availability. Some possibilities are straw, grass clippings, manure, wood chips, pine needles, peanut shells, etc. Seaweed is an especially good choice, as it adds a high concentration of minerals as it decomposes. Regardless of the kind of mulch you use, the reduction of water evaporation conserves both water and your energy and time spent watering.

Additional advantages of mulching include fewer weeds, lower incidence of soil-borne diseases, and deterrence of slugs and snails when a scratchy material, such as straw, is used. You will have cleaner fruit to harvest, especially if growing bush-type tomato plants, and soil erosion will not be a problem because of water runoff. A mulch can also act as a support for unstaked tomatoes, keeping the fruit off the soil.

Do not apply a mulch to your tomato plants until after plants are in bloom and have begun to set fruit. Mulching too early may inhibit vigorous growth. The material should always be put over damp rather than dry soil. You can also mulch your soil through the winter to protect against soil erosion and damage from severe freezing.

Black plastic sheeting is a popular mulching material. It absorbs the sun's heat, keeping the ground warm, and can be reused each year. It does not, however, add any nutrients to the soil. It can be used initially to warm up the soil and then be removed and replaced with an organic mulch which will gradually decompose. Some gardeners also feel that plastic does not allow the soil to breathe adequately, thus causing a souring of the soil.

STAKING

Whether or not to stake tomato plants is a controversial topic among gardeners. In general it can be said that early season varieties, or determinates, do not perform as well when staked and should be grown without support. Indeterminates or maincrop types which generally bear larger fruit can benefit from staking because of their longer growth, often into early fall.

Advocates of staking tomato plants see the following advantages: As a rule, the staked plant will yield fruit more quickly because more sun can reach the fruit. The fruit is also more uniform in size, cleaner, less likely to suffer rot and easier to harvest. Staked tomato plants take up less room than the bush or sprawling varieties, so your garden space is better utilized. Those who prefer to leave their plants unstaked believe harvests are larger, and there is less cracking, blossom-end rot and sunscald. (Often these same gardeners use a thick mulch to support the plants a little.) There is also considerably less labor for the gardener.

Tomatoes grown on stakes require training and pruning. You can choose to grow single-stem, double-stem or multiple-stem plants. In the first, all side shoots are pinched off as soon as they begin to form in the axils of the leaves, and only a single main stem is allowed to mature. A double-stem method is one in which a single side shoot is allowed to grow into a bearing stem. The multiple-stem method is when more than one side shoot is left to develop, usually 2 or 3. Whichever method you choose, careful pruning is important or strength will be sapped by non fruit-producing suckers. The unwanted side shoots and suckers should be pinched back only after 2 leaves have formed; then pinch just these leaves so that some foliage remains to protect the fruit. I prefer the multiple-stem method as there is higher yield and less chance of sunscald and other problems caused by lack of shade from foliage.

If using a simple wooden or bamboo stake, insert it into the hole at the time of transplanting the seedlings. The stake should be about 5 to 6 feet tall and sunk to a depth of about 6 inches in the soil. As the plant grows, tie the main stem or stems at several points along the stake with twistems or soft twine or cloth. The tie should be fixed securely to the stake, but loosely around the stem with a figure-8 loop.

There are several ways to supply supports for your tomato plants other than with simple stakes. You can construct a sawhorse-type frame and tie a tomato plant to each of the 4 legs; place large posts at either end of a row, run large-mesh fencing between them and tie the plants to the fencing; construct a simple teepee shape with stakes and tie the plants to the supports; create a simple ladderlike trellis to train the plants upright without need for tying; or construct foot-high square or rectangular wooden frames with wooden slats or wire fencing spanning them for supporting and training the plants. Or perhaps you have an existing fence which can be used for support. Once you understand the principle of staking you can create a method that works best for you and your garden.

You can avoid the work of staking and pruning by using wire cages for support. Both determinates and indeterminates can be trained this way. The fruit is lifted off the ground, the chance of sunscald is lessened and there is no need for tying the stems as the plants grow. Most indeterminates need cages approximately 5 feet high and 2 feet in circumference; determinates generally require the same unless they are a low-growing bush variety which can be accommodated by a shorter cage. Galvanized fence wire is an easy-to-use, low-maintenance material for constructing cages and it will last several years. Select wire with 4- to 6-inch squares to permit easy harvesting of fruit. Bend the wire into a cylinder, place it over the plant, and secure it in place with wickets or by removing the bottom rung of wire and forcing the vertical wire shafts into the soil.

GROWING IN CONTAINERS

If you have a limited amount of garden space or no soil at all, container gardening is an excellent way to produce an abundance of vine-ripened tomatoes for your kitchen. Clay or wood containers are best because they are porous and permit air to circulate, though plastic and ceramic may also be used. The size and shape of the container can vary from 50-gallon half wine barrels to foot-wide rectangular planter boxes to 12-inch clay pots. Your major concerns are adequate moisture and nutrients, both more critical in container gardening where the growing medium is limited in volume.

The soil used in containers must be light to permit good drainage and air circulation. It is best to use one of the commercial potting soil mixes or a mix of your own of half vermiculite and half peat moss. You will also need to fertilize the plants on a regular basis with a 5-10-10 formula; the schedule will depend on the size of the container and frequency of watering, a smaller container requiring more frequent applications. If you want to avoid this schedule, the new slow-release fertilizers can be applied at the time of transplant; nutrients are then released automatically throughout the growth period of the plant.

An advantage to container growing is that it is much easier to give the plants the amount of sunlight they need for healthy growth. Simply lift the containers and move them along the path of the sun. If your containers are not as mobile as this, consider casters or a dolly to make the move easier.

Some of the smaller tomato varieties, such as the cherry tomatoes, make excellent candidates for hanging containers. They can be trained to trail down the sides of containers and used as ornamentals on a deck, patio or walkway.

Container-grown tomatoes are staked according to the same principles as garden grown (see Staking), and will suffer the same pest and disease problems, particularly those caused by lack of moisture and nutrients if rigorous watering and fertilizing practices are neglected.

HARVESTING TOMATOES

Harvest tomatoes when fully ripe and firm. The last stages of ripening increase the sugar content and it is this that gives the tomato its flavor. It is no wonder home-grown tomatoes when picked fully ripe are superior to "store-bought" ones which are picked for shipping during the unripe stage. If possible, eat tomatoes promptly after harvesting when they have the fullest flavor. If you are harvesting your tomatoes while they are still green, pick them when they are of mature size but there is no sign of pink.

If your tomato plants are still producing fruit when fall comes and the weather has not turned too cold, cover the plants with burlap, bushel baskets, blankets or plastic at dusk. This will protect them against the cool night temperatures and give you a few more weeks of vine-ripened tomatoes. Or you may use the fiber glass or plastic protectors that protected the young transplants when they were first set out. Toward the end of the season some gardeners lay most of the stem of the plant along the ground, cover it with soil so roots form and start a new crop of tomatoes on the stem end that remains upright.

When frost threatens, pull up the tomato plants, including roots, and hang them upside down in a light warm place, such as a garage. Then pick the fruit as they ripen. An alternate method is to pick the fruit green off the vine and lay them in a single layer in shallow boxes or trays in a light warm place (60° to 75°F) to ripen.

TOMATO TROUBLES

There is no single variety of tomato that is free from pest and disease problems. Weather conditions, soil, air circulation, watering and feeding all play important roles in the growth of your tomatoes. Choose hybrid, disease-resistant varieties. Hybrid seeds and plants cost more, but the results are generally worth it: good yield and high-quality fruit. The vines are strong and husky, the flavor good and the plants produce profusely under the right conditions. When buying seeds or plants look for the letters "VF" or "VFN." This designates resistance to verticillium and fusarium wilt and root nematodes, the diseases most commonly suffered by tomato plants and all of which are fatal. Their symptoms include stunted growth, yellowing of the leaves and gradual defoliation of the plant.

A sunny, fertile, well drained and maintained tomato bed will not be subject to pests and diseases as a rule. A daily watchful eye should rid your garden of slugs, snails, tomato fruitworms, hornworms and caterpillars. Just pick them off. Planting marigolds, calendulas, garlic or chives amidst your tomato plantings will also discourage pests. A good hosing (early on a hot day) under the leaves will take care of aphids and a bad attack of white fly. Or place an aluminum foil "mulch" around the base of the plants to deter aphids; the sun's reflection will shoo them away. It is also important to never plant tomatoes in soil where plants of the same family have been grown (see Where to Plant).

Additional problems can occur. Blossom-end rot which results in a brown scar forming on the blossom end of the fruit is generally caused by one of two conditions: the soil drying out and then being heavily watered to compensate, or a lack of calcium in the soil which can be corrected by the addition of lime. Misshapen fruit can be caused by the tomatoes forming in temperatures too low for proper fertilization, in insufficient light, during rainstorms or because of excessive nitrogen in the soil. Cracked fruit may be the result of rain showers during extremely warm temperatures, thus encouraging too-rapid growth, or too much direct sun striking the tomato. Sunscald is a common problem when foliage is not sufficient to protect fruit from the penetrating rays of the sun. Yellowish patches which eventually form blisters appear on green, immature fruit. Catfacing is the term used to describe puckered areas on the fruit caused by cool weather during bloom. The blossoms adhere to the tiny, immature fruit and as the fruit grows it gradually becomes malformed.

Remember that the best defense against most "tomato troubles" is a clean, well-maintained tomato bed.

soups & drinks

CREAMED TOMATO SOUP

1 carrot, diced
1 celery rib with leaves, diced
2 tablespoons butter
1 tablespoon flour
2 pounds tomatoes, peeled, seeded and chopped
1 garlic clove, minced
1/2 teaspoon dried thyme
1/2 teaspoon sugar
1-1/2 quarts beef or chicken stock
1/2 cup heavy cream
salt and freshly ground pepper to taste

In a soup pot sauté carrot and celery in butter over medium heat for 5 minutes. Stir in flour and cook and stir 2 minutes. Add tomatoes, garlic, thyme, sugar and stock. Bring to a rapid boil, then lower heat, cover and simmer 1 hour. Purée mixture and return to heat. Stir in cream and just heat through, about 2 minutes; do not bring to a boil. Season with salt and pepper and serve immediately.
Serves 8

TOMATO AND RICE SOUP

2 leeks, including green tops, chopped
1 small onion, chopped
2 tablespoons butter
2 pounds tomatoes, peeled, seeded and chopped
2 garlic cloves, minced
4 cups chicken stock
1/2 teaspoon sugar
bouquet garni of 4 parsley sprigs, 6 fennel seeds, 1 bay leaf, 1 thyme sprig
1/4 cup long-grain rice
salt and freshly ground pepper to taste
chopped parsley or basil

In a large soup pot, sauté the leeks and onion in butter until onion is translucent. Add the tomatoes, garlic, stock and sugar. Add bouquet garni and bring to a rapid boil; reduce heat to moderate and cook for 10 minutes. Add rice and continue cooking for 25 minutes or until rice is tender. Discard bouquet garni, season with salt and pepper and serve.
Serves 6 to 8

TOMATO BOUILLON

1 pound Italian plum tomatoes, peeled, seeded and chopped
4 cups chicken stock
1/4 cup vermouth or dry white wine
1 tablespoon sugar
salt and freshly ground pepper to taste
garnish: thin shreds of lemon peel

Combine the tomatoes and stock and bring to a boil. Reduce heat and simmer 15 minutes. Put through a fine sieve and season broth with vermouth, sugar, salt and pepper. Reheat, and serve in soup mugs with garnish. May also be served cold.
Serves 4 to 6

SEAFOOD GUMBO

2 onions, chopped
1 garlic clove, minced
3 tablespoons butter
2 tablespoons flour
1 pound tomatoes, peeled
  and chopped
1 pound okra, sliced
pinch cayenne pepper
1-1/2 quarts fish stock or water

1/2 cup dry white wine
1 bay leaf
1 pound firm fish fillets, cut in
  1-inch chunks
1 pound raw shrimp, shelled and
  deveined
1 pint oysters with liquor
salt and freshly ground pepper to taste
6 to 8 cups freshly cooked rice

In a soup kettle sauté the onions and garlic in butter until translucent; add the flour and cook slowly, stirring, for 2 minutes. Add tomatoes, okra, cayenne pepper, stock, wine and bay leaf. Bring to a rapid boil, lower heat, cover and simmer 45 minutes. Add the fish, shrimp and oysters and their liquor and cook 15 minutes more. Season with salt and pepper to taste. Serve in soup plates over rice. Pass the Tabasco or Worcestershire sauce, if desired.
Serves 6 to 8

TOMATO AND ONION SOUP

2 pounds onions, thinly sliced
1 garlic clove, minced
2 tablespoons each olive oil and butter
3 pounds tomatoes, peeled, seeded
  and chopped
1 bay leaf

1-1/2 quarts water
3 tablespoons minced parsley
salt and freshly ground pepper to taste
accompaniments:
  grated Parmesan cheese
  French bread and sweet butter

In a soup pot sauté the onions and garlic in oil and butter until translucent. Add the tomatoes and bay leaf and simmer 10 minutes. Add the water, bring to a rapid boil, then lower heat and simmer 10 minutes. Remove bay leaf. Purée mixture; return to heat and heat through. Stir in parsley, season with salt and pepper and serve with accompaniments.

Serves 8

Variations
- Place a poached egg on the bottom of each soup bowl. Ladle soup over.
- Add 1/2 cup dry white wine and 1 teaspoon dried tarragon or basil to purée.
- Serve with a dollop of sour cream and snipped chives.

TOMATO-VEGETABLE SOUP WITH BOILED CHICKEN

1 3-1/2- to 4-pound whole chicken
1-1/2 pounds tomatoes, peeled, seeded and chopped
2 onions, chopped
2 carrots, chopped
2 celery ribs with leaves, chopped
2 turnips, chopped
2 potatoes, peeled and chopped
2 leeks, chopped
bouquet garni of 2 parsley sprigs, 1 rosemary sprig and 1 bay leaf
2 quarts water
salt and freshly ground pepper to taste
accompaniments:
    toasted croûtons
    grated Gruyère cheese

Combine all the ingredients in a large soup pot and bring to a rapid boil. Skim off any surface scum and simmer, covered, for 1 hour. Remove chicken to a platter and keep warm. Discard bouquet garni. Adjust seasonings and serve in large soup bowls accompanied with croûtons and grated cheese. Serve chicken as a separate course.

Serves 6

MANHATTAN CLAM CHOWDER

4 slices bacon, diced
2 onions, chopped
2 carrots, chopped
2 celery ribs with some leaves, chopped
3 tablespoons chopped parsley
1/2 teaspoon dried thyme
1 bay leaf
1 pound potatoes, peeled and diced
2 pounds tomatoes, peeled, seeded and chopped
1-1/2 quarts water
1 pint clams with liquor
salt and freshly ground pepper to taste

Sauté bacon until fat is released; add onions and cook until translucent. Add remaining ingredients, except clams and salt and pepper, with 1-1/2 quarts of water. Bring to a rapid boil. Skim any surface scum. Cover and simmer for 40 minutes. Coarsely chop the clams, add to the soup with their liquor and simmer 5 minutes. Season with salt and pepper.
Serves 6

ICED TOMATO-LIME CREAM SOUP

1 medium onion, chopped
2 tablespoons butter
1-1/2 pounds tomatoes, peeled, seeded and chopped
1/2 cup chicken stock
1/2 teaspoon sugar
1/2 teaspoon dried thyme
1/2 teaspoon salt
3/4 cup heavy cream
1/4 cup sour cream
2 tablespoons fresh lime juice
1 tablespoon grated lime peel
salt and freshly ground pepper to taste
garnish: thin lime slices and parsley sprigs

Sauté the onion in butter until translucent. Add tomatoes, chicken stock, sugar, thyme and salt and simmer for 15 minutes. Let mixture cool and purée it, then put it through a fine sieve. Blend mixture well with both creams, lime juice, grated peel and salt and pepper to taste. Pour into 6 small glass bowls and chill in freezing compartment of refrigerator for 1 hour. Garnish each bowl with a slice of lime and sprigs of parsley.
Serves 6

GAZPACHO

Soup Base
1/2 cup fresh French bread crumbs
2 garlic cloves
1/4 cup olive oil
1 onion, sliced
1 cucumber, peeled, seeded and cut up
1 green pepper, cut up
2 pounds tomatoes, peeled and cut up
1/4 cup red wine vinegar
1 cup each Canned Tomato Juice, page 87, and water
pinch of cayenne pepper
1/4 teaspoon ground cumin
1/2 teaspoon salt
freshly ground pepper to taste

Garnish
1 small sweet red or green pepper, finely chopped
1 small cucumber, chopped
1 small onion, chopped
2 medium tomatoes, peeled, seeded and chopped
1-1/2 cups toasted croûtons
chopped coriander (optional)

Combine all the ingredients for soup base, place in a blender in several batches and purée. Pour into a bowl and chill for at least 3 hours. Prepare all garnish ingredients and place each ingredient in a separate mound on a platter. When ready to serve, let each guest add his own garnishes to his bowl of soup.
Serves 8

QUICK TOMATO JUICE

Put red ripe juicy tomatoes through a juicer. Season with a pinch of sugar, salt and freshly ground pepper, and fresh lemon juice to taste.

TOMATO JUICE BEVERAGES

The following drinks are to be made with 1 cup Canned Tomato Juice, page 87, or Quick Tomato Juice, preceding, or commercially canned juice may be used.

Tomato Juice Brittania Add 2 watercress sprigs, dash Worcestershire sauce and a few grinds of pepper with the tomato juice to a blender and blend 30 seconds. Pour over ice.
Tomato Juice Napoli Add 3 or 4 fresh basil leaves, 1 thin onion slice and salt and freshly ground pepper with the tomato juice to a blender and blend 30 seconds.
Tomato Juice Italienne Prepare as for Tomato Juice Napoli, preceding, adding 1/2 garlic clove and 1/2 teaspoon fresh lemon juice.
Tomato Juice Mexicali Add 1 thin slice fresh green chili pepper, 1 coriander (cilantro) sprig, a pinch oregano and 2 avocado slices with the tomato juice to a blender and blend 30 seconds.
Tomato Juice Creole Add 1 green pepper strip, dash each fresh lime juice and Tabasco sauce, 1/2 garlic clove and salt and freshly ground pepper with the tomato juice to a blender and blend 30 seconds.
Tomato Juice San Joaquin Add 1/2 cup sliced carrots, 1 parsley sprig, 1 tablespoon fresh lemon juice, dash Worcestershire sauce and salt and freshly ground pepper with the tomato juice to a blender and blend 1 minute. Pour over ice and top with a dollop of sour cream.
Tomato Juice Fresno Add 1/2 cup chopped celery with tomato juice to a blender and blend 1 minute. Pour over ice and garnish with a slice of lime.

Tomato Juice St. Helena Add 4 thin peeled cucumber slices, 1 thin onion slice and a small dill sprig with tomato juice to a blender and blend 30 seconds.

Tomato Juice Marrakech Add 1/4 cup yoghurt, 1/2 teaspoon each Worcestershire sauce and fresh lemon or lime juice, and a dash of Tabasco sauce with the tomato juice to a blender and blend 30 seconds. Serve with a celery, cucumber or fennel stick as stirrer.

BLOODY MARY

1-1/2 ounces vodka
1/2 cup tomato juice
1 tablespoon fresh lemon or lime juice

dash each Worcestershire sauce and Tabasco sauce
salt and freshly ground pepper to taste
garnish: cucumber or celery stick

Combine all the above ingredients and pour over ice in a tall glass. Serve with garnish as stirrers.
Serves 1

Bloody Maria Prepare as for Bloody Mary, omitting vodka and using 1-1/2 ounces tequila.

Bloody Marie Prepare as for Bloody Mary, reducing lemon juice to 1 teaspoon and adding 1 teaspoon pernod.

Bloody Shanghai Prepare as for Bloody Mary, reducing lemon juice to 1 teaspoon, omitting Worcestershire sauce, salt and pepper and garnish, and adding 1/2 teaspoon soy sauce and 1/2 cup crushed ice. Combine all ingredients in a cocktail shaker, shake well and strain into a glass.

Bloody Bullshot Prepare as for Bloody Mary, reducing tomato juice to 1/4 cup and adding 1/4 cup beef bouillon, chilled, and 1/2 cup crushed ice. Combine all ingredients in a cocktail shaker, shake well and strain into a glass; omit garnish.

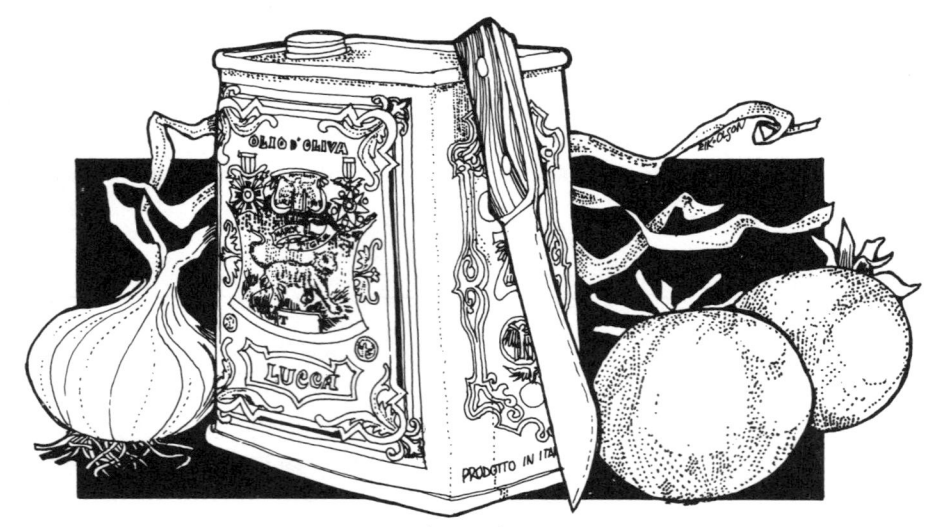

salads

TOMATO SALAD DRESSING

1 cup Canned Tomato Juice, page 87
2 tablespoons wine vinegar or fresh lemon juice
1 garlic clove, crushed
1 teaspoon each chopped chives and parsley

1/4 teaspoon dry mustard
2 teaspoons sugar
1/2 teaspoon salt
1/2 teaspoon freshly ground pepper

Put all the ingredients into a screw top jar and shake well. Use as a dressing for salad greens. Also good on cold seafood.
Makes about 1 cup

VINAIGRETTE DRESSINGS

Use 4 or 5 parts good vegetable salad oil such as corn, safflower, peanut, sesame (not the Oriental variety), walnut or olive oil to 1 part good wine vinegar or fresh lemon juice. Add salt, freshly ground pepper and Dijon-style mustard to taste. Fresh or dried herbs may be added if desired.

Basic Vinaigrette
1/2 cup salad oil (preferably olive or walnut oil)
1/2 teaspoon salt
1/2 teaspoon freshly ground pepper
1/2 teaspoon Dijon-style mustard (optional)
2 tablespoons wine vinegar or fresh lemon juice or to taste (depending on vinegar strength)

Combine the oil, salt, pepper and mustard in a bowl and mix well, then gradually add vinegar to desired taste.
Makes about 1/2 cup

With Garlic Add 1 garlic clove, minced or pressed through a garlic press.
With Blue or Roquefort Cheese Add 3 tablespoons crumbled cheese, mixing well.
With Herbs Add to taste any one of the following herbs: chopped dill, chervil, tarragon, basil, mint or coriander. Chives, green onions or parsley may be added with any of the herbs.
With Tomato Catsup Add 2 tablespoons Tomato Catsup, page 86, to Basic or Herb Vinaigrette, mixing well.

TOMATO SALADS

Fresh, fully vine-ripened tomatoes may be stuffed, sliced or cut up in a multitude of salads. Tomatoes should be served at room temperature for best flavor. Serve on a bed of any kind of lettuce, spinach, watercress, alfalfa sprouts, finely shredded Chinese cabbage or head cabbage, and garnish with olives, nuts or chopped parsley, chives, green onion or dill.

Stuffed Tomatoes Prepare tomatoes for stuffing, page 6, and fill just before serving with any of the following stuffings. Chill stuffing, if desired, but not the tomatoes. Toss stuffings with your favorite dressing.
- potato, macaroni, rice, bean or lentil salad
- chicken, turkey or ham salad
- cottage cheese
- diced Swiss, cheddar, Jarlsberg or Monterey jack cheese
- shrimp, crab or lobster salad
- cooked and diced roast beef, lamb, pork or veal
- flaked tuna, salmon or other cooked fish
- egg salad
- cooked and diced vegetables such as beets, carrots, potatoes, celery root, cauliflower, artichoke hearts or bottoms, broccoli, green beans or peas, corn, asparagus or lima beans
- cole slaw or carrot salad
- grated, diced, julienne-cut or sliced fresh vegetables such as carrots, celery, cucumber, sweet red or green pepper, radishes, fennel or mushrooms

Sliced Tomatoes If the skin is thin, tomatoes for slicing do not need to be peeled. Tomatoes may also be chopped or cut in wedges for salads. Unpeeled, sliced or cut-up tomatillos may be used instead of tomatoes in any of the following ways.
- Serve unseasoned, or with salt and freshly ground pepper.
- Sprinkle with chopped chives, green onions, Bermuda or Spanish onions, garlic or shallots. Dress with Vinaigrette Dressing, page 33, or dressing of choice.
- Sprinkle with chopped fresh herbs such as parsley, basil, coriander, dill, mint, tarragon or chervil.
- Sprinkle with fresh lemon or lime juice and a little sugar, salt and freshly ground pepper.
- Top with a dollop of sour cream, mayonnaise or plain yoghurt before serving.
- Sprinkle with salt and freshly ground pepper and a dash of cayenne pepper; pour over Madeira wine to cover and let stand 1 hour before serving.
- Top with crisp bacon, crumbled if desired (especially good on a bed of crisp, young spinach leaves).
- Top with crumbled feta or blue cheese or grated Swiss or sharp cheddar cheese.
- Serve with cottage cheese.
- Serve with anchovy fillets, sardines, kippered herring or flaked tuna.
- Serve with sliced raw zucchini, mushrooms or avocado and dress with Vinaigrette Dressing, page 33.
- Sprinkle with chopped toasted or raw nuts or sunflower, pumpkin or sesame seeds.
- Layer with thin slices of prosciutto.
- Combine with cucumber, red onion and radish slices and dress with Vinaigrette Dressing, page 33, or dressing of choice.

MARINATED TOMATOES

Firm, ripe tomatoes may be cut in slices, wedges or chunks for marinating. Serve as an appetizer or first course, as a main dish accompaniment, on greens or as salad tossed with greens or vegetables. Following are some marinade suggestions for 1-1/2 to 2 pounds tomatoes. Tomatillos or whole cherry tomatoes may also be used.

BASIC MARINADE

1/2 cup corn oil or olive oil
2 tablespoons red or white wine vinegar
1/2 teaspoon each sugar and salt
1/4 teaspoon freshly ground black or white pepper

Combine all ingredients and pour over tomatoes in a serving dish; let stand 30 minutes at room temperature.

Blue-Cheese Variation Add 1/4 cup crumbled or grated blue cheese (freeze before grating) to Basic Marinade.
Mixed-Herbs Variation Add 1/2 teaspoon dried tarragon, 2 tablespoons minced parsley and 1 tablespoon minced chives to Basic Marinade.
Bermuda-Onion Variation Add 1 large Bermuda onion, cut in thin slices, to Basic Marinade.
Mushroom Variation Add 1 cup sliced mushrooms to Basic Marinade.

TOMATO FRAPPÉ

2 cups peeled, seeded and chopped tomatoes
6 peppercorns
1 bay leaf

2 whole cloves
2 tablespoons sugar
1 teaspoon salt
1 lemon slice

Combine all the ingredients in a saucepan and simmer for 15 minutes. Rub through a sieve and pour sieved mixture into a freezer tray. Freeze partially, about 20 minutes, then stir well with a spoon. Freeze another 20 minutes and stir again. Freeze until firm. Fill sherbet glasses and serve as an accompaniment to meat, fish or poultry.
Serves 4

ZUCCHINI AND TOMATO SALAD

4 to 6 small zucchini, sliced on the diagonal 1/4 inch thick
olive oil
4 to 6 firm tomatoes, sliced 1/4 inch thick

1 tablespoon chopped thyme
salt and freshly ground pepper to taste
1/2 cup Vinaigrette Dressing, page 33

Sauté the zucchini in olive oil until lightly browned on both sides. Remove with slotted spoon and drain on paper toweling. When zucchini are cool, alternate layers of zucchini and tomato slices on a platter. Sprinkle with thyme, salt and pepper and pour Vinaigrette Dressing over all.
Serves 4 to 6

TOMATO SALATA

4 large firm, ripe tomatoes, peeled,
  seeded and cut in chunks
4 green peppers, cut in 1-inch chunks
1 onion, quartered and sliced
  crosswise 1/4 inch thick
1 2-ounce can anchovy fillets, drained
2 tablespoons capers
1/4 cup minced Italian parsley
1 garlic clove, minced
1/2 cup olive oil
3 tablespoons wine vinegar
salt and freshly ground pepper to taste
Romaine lettuce leaves

Combine all ingredients except lettuce and mix lightly. Chill for several hours. Serve on lettuce leaves.
Serves 6

ARMENIAN SALAD

4 large firm, ripe tomatoes, peeled,
  seeded and diced
2 cucumbers, peeled, seeded and diced
2 celery ribs, diced
1 small red onion, chopped
1/4 cup chopped coriander or parsley
1 cup chopped watercress
1 tablespoon chopped mint
1 tablespoon chopped basil
1/4 cup olive oil
1/4 cup fresh lemon juice
salt and freshly ground pepper to taste

Combine all the vegetables and herbs. Sprinkle with olive oil, lemon juice, salt and pepper. Toss and serve immediately.
Serves 6

GUACAMOLE SALAD

3 ripe avocados, mashed
1/3 cup fresh lemon juice
2 tablespoons grated onion
1 garlic clove, minced
1 fresh green chili pepper, seeded and minced, or
1/8 to 1/4 teaspoon cayenne pepper
salt and freshly ground pepper to taste
lettuce leaves
4 tomatoes, peeled, seeded and chopped
1 cucumber, peeled, seeded and chopped
1 green pepper, chopped
Tortilla Chips, following

Combine the avocados, lemon juice, onion, garlic, chili pepper, salt and pepper. Arrange lettuce leaves on a platter. Mound avocado mixture in the center and surround with the tomatoes, cucumber and green pepper. Serve with Tortilla Chips. Serves 6

Tortilla Chips Cut stale corn tortillas in pie wedges, about 6 pieces per tortilla, and fry in lard or vegetable oil until crisp. Drain and sprinkle lightly with salt.

ITALIAN SALAD

1/2 pound green beans, cooked and cut in 1-inch pieces
2 cups diced cooked potatoes
1-1/2 pounds tomatoes, seeded and diced
1 7-ounce can tuna fish, drained and flaked
2 tablespoons chopped parsley
2 tablespoons chopped basil
1/4 cup chopped green onions
1/2 cup Vinaigrette Dressing, page 33
Romaine lettuce leaves

Combine green beans, potatoes, tomatoes, tuna fish, parsley, basil, green onions and Vinaigrette Dressing and chill at least 30 minutes. Serve on a bed of lettuce.
Serves 4

TOMATO FANS

6 medium firm, ripe tomatoes
lettuce leaves or watercress sprigs
1 pound cooked baby shrimp or flaked crab meat
1 egg yolk
1/4 cup wine vinegar
1 teaspoon Dijon-style mustard
1/2 cup olive oil
2 tablespoons each minced chives and parsley
1 tablespoon minced dill
salt and freshly ground pepper to taste

Cut each tomato in 6 wedges almost to stem end and fan out wedges. Place tomatoes on lettuce leaves or beds of watercress sprigs; heap with shrimp or crab meat. Whisk together egg yolk, wine vinegar and mustard. Gradually add oil in a slow stream while beating, and continue beating until thick. Add herbs and salt and pepper and pour over filled tomatoes.
Serves 6

STUFFED TOMATOES WITH MOZZARELLA AND HAM SALAD

4 large firm, ripe tomatoes
lettuce leaves
1/4 pound grated mozzarella cheese
1/4 pound cooked ham, cut in julienne
3 tablespoons chopped basil

1/4 cup olive oil
1 tablespoon fresh lemon juice or wine vinegar
salt and freshly ground pepper to taste

Prepare tomato shells for stuffing, page 6, or cut each tomato in 6 wedges almost to stem end and fan out wedges. Place tomatoes on lettuce leaves and fill with cheese and ham. Combine the basil, olive oil and lemon juice or vinegar and drizzle over filled tomatoes. Season with salt and pepper.
Serves 4

STUFFED TOMATOES WITH TURKEY SALAD

6 large firm, ripe tomatoes
lettuce leaves or watercress sprigs
3 cups diced cooked turkey
1-1/2 cups thinly sliced celery with some leaves
2 tablespoons minced green onions

1/4 cup mayonnaise
1/4 cup sour cream
salt and freshly ground pepper to taste
1/4 cup toasted slivered almonds or walnuts

Prepare tomato shells for stuffing, page 6, or cut each tomato in 6 wedges almost to stem end and fan out wedges. Place tomatoes on lettuce leaves or beds of watercress sprigs. Combine turkey, celery, minced onions, mayonnaise and sour cream, mixing well. Season with salt and pepper and heap salad into tomatoes; sprinkle nuts over.
Serves 6

STUFFED TOMATOES WITH EGGS AND CHEESE

4 large firm, ripe tomatoes
lettuce leaves
4 hard-cooked eggs, minced
1 cup grated Swiss cheese
1/2 cup mayonnaise or sour cream

2 tablespoons chopped ripe olives
1 tablespoon prepared mustard
1 teaspoon vinegar
freshly ground pepper to taste

Prepare tomato shells for stuffing, page 6, or cut each tomato in 6 wedges almost to stem end and fan out wedges. Place tomatoes on lettuce leaves. Combine all remaining ingredients, mixing lightly. Heap salad into tomatoes.
Serves 4

TOMATO ASPIC

3-1/2 cups peeled, seeded and chopped
    tomatoes
1 teaspoon salt
2 teaspoons sugar
1/2 teaspoon paprika
2 tablespoons fresh lemon juice
1/4 cup chopped onion

3 celery ribs with leaves
1 bay leaf
dash Tabasco sauce
2 scant tablespoons (2 packages)
    unflavored gelatin, dissolved in
1/2 cup cold water

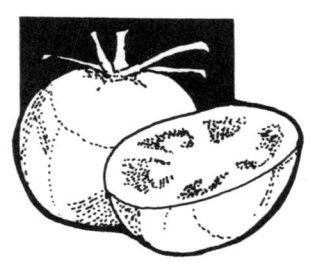

Put all the ingredients except gelatin in a heavy saucepan and boil, covered, for 30 minutes. Strain through a fine sieve and measure out 3-1/2 cups of liquid. Blend in gelatin mixture until dissolved. Pour into a wet 4-cup mold, let cool and chill until set. Unmold by inverting on a chilled plate and covering for a few seconds with a cloth that has been soaked in hot water and wrung out.
Serves 6 to 8

Variations Using a 6- to 8-cup mold, add 1 to 2 cups of any of the following ingredients to the aspic after it is partially set; then chill until firm: sliced olives, diced or chopped cucumber, celery, or green pepper, grated or chopped carrots or radishes, diced cooked meats, chicken or turkey or cooked baby shrimp, clams, oysters or flaked fish (such as tuna).

QUICK TOMATO ASPIC

2 scant tablespoons (2 packages) unflavored gelatin, dissolved in 1/2 cup cold Canned Tomato Juice, page 87

3-1/2 cups Canned Tomato Juice, page 87, heated

Combine gelatin mixture with the hot tomato juice. Blend well and pour into a wet 4-cup mold; cool, then chill until set. May be used with above variations.

TOMATO ASPIC FREEZE

Prepare Quick Tomato Aspic, preceding. Let cool and freeze partially in freezer trays. Beat 2 egg whites until stiff and fold into partially frozen aspic. Return to freezer tray and chill until firm. Serve in sherbet glasses with a dollop of sour cream or mayonnaise as a first course or side accompaniment to meat, seafood or poultry.

hot side dishes

FRIED TOMATO SLICES

Use green or firm, ripe red tomatoes, cut in 1/2-inch thick slices. Dip into cornmeal seasoned with salt and pepper to coat lightly and fry in hot bacon drippings or half butter and half corn oil until crisp. Drain on paper toweling. Serve immediately. Good served with crisp bacon and eggs for breakfast.

NEAPOLITAN FRIED RED OR GREEN TOMATOES

2 pounds firm red or green tomatoes, sliced 1/4 inch thick
1/2 teaspoon dried oregano
salt and freshly ground pepper to taste
2 eggs, beaten
1/2 cup milk
1 cup flour
1-1/2 cups fine dry bread crumbs
olive oil
grated Parmesan cheese (optional)

Sprinkle tomato slices with oregano and salt and pepper. In a bowl, beat together the eggs and milk. Put the flour and bread crumbs in separate dishes. In a large skillet, heat 1/2-inch of oil until very hot. Dip each tomato slice in flour, coating both sides, then in the egg-milk mixture, then in bread crumbs, coating well on all sides. Fry in single layer in the hot oil for about 1 minute on each side or until lightly browned. Drain on paper toweling and sprinkle with Parmesan cheese, if desired. Serve immediately.
Serves 4 to 6

BAKED SLICED TOMATOES

8 tomatoes, sliced 1/2 inch thick
2 tablespoons butter
1 onion, thinly sliced
1 tablespoon brown sugar
1 tablespoon chopped dill
1/2 cup dry bread crumbs
1/2 cup grated Parmesan cheese

Place a layer of half the tomato slices in a buttered shallow baking dish, dot with butter and top with a layer of half the onion slices. Repeat layers. Sprinkle remaining ingredients over top and bake in a preheated 350° oven for 30 minutes or until golden.
Serves 4 to 6

BROILED TOMATO HALVES

Use firm ripe tomatoes. Cut in half horizontally, seed if desired, and place on a baking sheet.

With Herbs Brush cut side with olive oil or melted butter. Season with salt and freshly ground pepper and a pinch of any one or two of the following herbs: basil, chervil, chives, mint or dill, marjoram or oregano, thyme, rosemary, coriander or parsley. Broil under moderate heat until edges are singed; do not overcook.

With Crumbs and Cheese Combine equal parts of dry bread crumbs and grated Parmesan or cheddar cheese, sprinkle on tomatoes and broil on lowest rack 5 to 10 minutes. Watch carefully to prevent scorching.

With Sour Cream Combine equal parts sour cream and mayonnaise and season with curry powder, salt and freshly ground pepper to taste. Spread about 2 tablespoons of mixture on top of each tomato half and broil 5 to 10 minutes under heat until bubbly.

BAKED CHEDDAR TOMATOES

2 pounds tomatoes, peeled and sliced 1/2-inch thick
2 tablespoons dry sherry

salt and freshly ground pepper to taste
1 cup grated cheddar cheese
minced parsley for garnish

Place tomato slices in a buttered shallow baking dish. Drizzle with sherry, salt and pepper lightly and bake in a preheated 300° oven for 20 minutes. Sprinkle cheese over the tomatoes and continue to bake for 15 minutes more. Garnish with parsley. Serves 4 to 6

SCALLOPED TOMATOES

1 cup coarse dry bread crumbs
1 teaspoon sugar
1/2 teaspoon salt
1/4 teaspoon freshly ground pepper
4 tablespoons butter
2 pounds tomatoes, sliced

Combine the crumbs, sugar, salt, pepper and butter. Put a layer of tomatoes in a shallow baking dish and sprinkle some of the crumb mixture on top. Repeat layers ending with a crumb topping. Bake in a preheated 350° oven for 30 minutes.
Serves 4

Variation Reduce butter measure to 2 tablespoons and add 1/4 cup grated Parmesan cheese to bread crumb mixture.

BACON, TOMATO AND GREEN BEAN SAUTÉ

4 slices bacon, diced
1 onion, diced
2 teaspoons sweet Hungarian paprika
1 pound tomatoes, peeled and cut in chunks
1 pound green beans, cut on the diagonal in 1-inch pieces
1 teaspoon sugar
salt and freshly ground pepper to taste

In a large skillet sauté the bacon until fat begins to melt. Add onion and continue sautéing 5 minutes; stir in paprika and cook 1 more minute. Add tomatoes, green beans and sugar, combining well. Cover and cook 10 minutes or until beans are tender. Season with salt and pepper to taste.
Serves 4 to 6

SUCCOTASH

2 slices bacon, diced
1 green pepper, diced
1 pound lima beans, shelled
2 cups fresh corn kernels

1 pound tomatoes, peeled and diced
salt and freshly ground pepper to taste
1/4 teaspoon freshly grated nutmeg

Sauté the bacon in its own fat with the green pepper for 2 minutes. Add remaining ingredients. Cover and cook on low for 10 to 15 minutes or until vegetables are tender.
Serves 4

MIXED VEGETABLE SAUTÉ, ETHIOPIAN STYLE

1 pound potatoes, peeled and cut like French fries
2 onions, sliced
1/4 cup olive oil
1 pound green beans, cut in 2-inch pieces

1 teaspoon ground turmeric
1 pound tomatoes, peeled and chopped
2 garlic cloves, minced
salt and freshly ground pepper to taste

Put the potatoes in a bowl of cold water. Sauté onions in oil until they are translucent. Add the beans and turmeric and continue to sauté for 5 minutes longer. Add tomatoes and garlic and cook on high heat for 2 minutes. Drain the potatoes and add them to the pot. Cover and simmer for 15 minutes or until potatoes are tender. Season with salt and pepper.
Serves 4 to 6

TOMATO TIMBALES

4 shallots, minced
1/4 cup minced onion
4 tablespoons butter
2 pounds tomatoes, peeled, seeded and chopped
1 tablespoon Canned Tomato Paste, page 84
1 tablespoon sugar
1/2 teaspoon dried basil
6 egg yolks, beaten
3/4 cup heavy cream
1/8 teaspoon freshly grated nutmeg
1/2 teaspoon salt
1/4 teaspoon freshly ground pepper
garnish: 2 tablespoons minced parsley

In a large skillet, sauté the shallots and onion in butter until onion is translucent. Add tomatoes, tomato paste, sugar, basil and cook mixture over moderate heat, uncovered, for 30 minutes or until mixture is thickened. Force through a sieve into a bowl. Cool mixture. Add egg yolks, cream, nutmeg, salt and pepper and blend well. Pour into 6 buttered timbale molds or custard cups to within 1/2 inch of top. Cover with buttered waxed paper. Set timbales in a baking pan and pour enough boiling water to reach two-thirds up sides of molds. Bake in a preheated 350° oven for 35 minutes or until set. Remove from water and let stand 5 minutes. Run a sharp knife around sides of timbales to loosen and unmold onto a platter. Garnish with parsley. Serves 6

SAUTÉED CHERRY TOMATOES

Sauté whole cherry tomatoes in butter for 2 or 3 minutes; add salt and freshly ground pepper to taste. Sprinkle with a little sugar to make them shine. Remove from heat and serve immediately.

STEWED TOMATOES

3 pounds tomatoes, peeled
3 tablespoons butter
1/2 teaspoon sugar
salt and freshly ground pepper to taste

Put the tomatoes, butter and sugar in a heavy saucepan, cover and cook on low until tomatoes are tender, about 15 minutes. Season with salt and pepper and serve.
Serves 6 to 8

With Onions Thinly slice a small onion and cook with the tomatoes.
With Garlic Mince a garlic clove and cook with the tomatoes.
With Fresh Basil Stir in 2 tablespoons chopped basil when ready to serve.
With Cinnamon Add a pinch of ground cinnamon with salt and pepper.
With Bread Crumbs Blend in 1/2 cup dry bread crumbs the last 5 minutes of cooking.

Stewed Tomatoes with Other Vegetables Halve ingredients for Stewed Tomatoes, preceding, add one of the following vegetables and proceed with recipe for Stewed Tomatoes.
- 1 pound zucchini, sliced, and 1/2 teaspoon dried oregano
- 1 large eggplant, unpeeled and cut in 1-inch cubes, with 1/2 teaspoon dried oregano and 1 garlic clove, minced
- 3 cups freshly cut corn kernels, 1 small green pepper, diced, and 1 small onion, diced
- 1 head cauliflower, separated into flowerets
- 1 pound green beans or Italian beans, cut in 2-inch lengths

TOMATO RICE PILAF

1 cup long-grain rice
4 tablespoons butter
1 onion, chopped
1 garlic clove, minced

2 cups peeled, seeded and chopped
   tomatoes
2 tablespoons chopped parsley
1-1/2 cups chicken stock

Sauté the rice in the butter until lightly golden over medium heat, stirring constantly. Add the onion and garlic and continue cooking 5 minutes. Add remaining ingredients, bring to a rapid boil and cook uncovered until bubbles disappear on surface and most of the liquid has been absorbed. Immediately lower heat to a simmer, cover and simmer 20 minutes or until rice is tender. Fluff with a fork and serve as an accompaniment to meat, seafood, poultry or eggs.
Serves 4

SPANISH RICE

3 tablespoons olive oil
1 onion, finely chopped
1 cup long-grain rice
2 cups chicken or beef stock or water

1 cup Canned Tomato Sauce, page 84
1/4 teaspoon ground cumin (optional)
salt and freshly ground pepper to taste

In a heavy skillet or saucepan, heat the oil and sauté onion until limp. Add the rice, stirring constantly until golden. Add remaining ingredients and simmer uncovered, without stirring, until rice is tender and all liquid is absorbed, about 20 minutes.
Serves 6

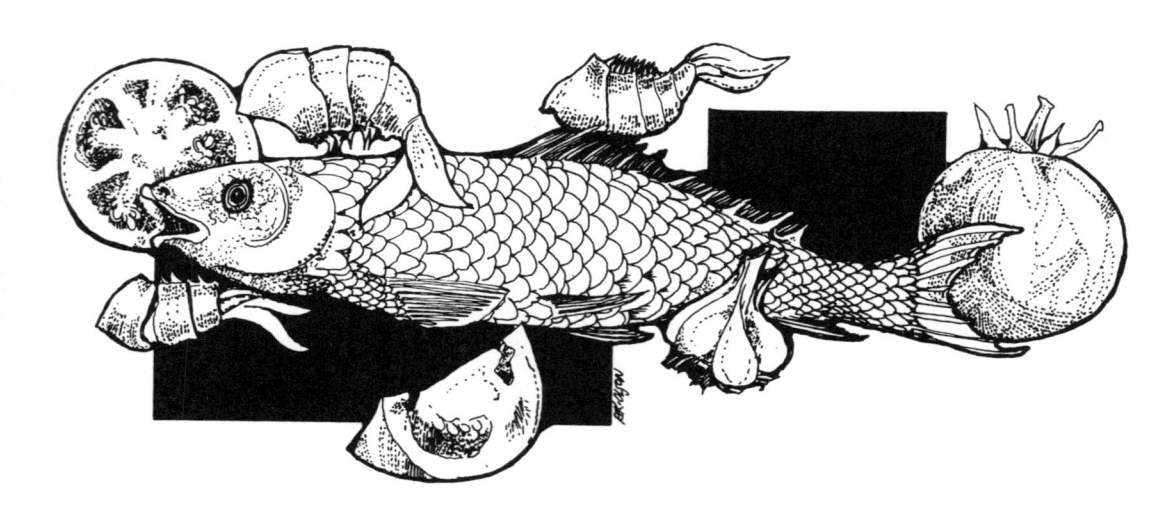

main dishes

TOMATO SANDWICHES

When preparing tomato sandwiches, always make them up just prior to serving to prevent the bread from getting soggy. Draining or seeding sliced tomatoes, and using only firm, ripe tomatoes will also prevent excess moisture. To make excellent sandwiches, begin with nutritious bread with good texture, such as whole-grain bread. For rolls, use the French or Italian type, bagels, onion, poppy or sesame seed buns. Good white bread made from unbleached flour or egg twist bread, plain or toasted, may be used. Fresh sweet butter is preferred to salted butter. Use high-quality mayonnaise (preferably homemade) and mustards, such as Dijon-style mustard. Unpeeled, sliced tomatillos may be substituted for tomatoes in any of the following recipes.

Tomato Sandwich Spread bread of your choice with butter or mayonnaise. Top with tomato slices, salt and pepper lightly if desired, and serve at once.

With Salad Greens Add crisp lettuce, spinach leaves or watercress sprigs to sandwich before topping with tomato slices.

With Bacon and Avocado Add a layer of sliced avocado and 3 or 4 strips of crisp bacon to sandwich before topping with tomato slices. (A thin slice of fried ham may be used instead of the bacon.)

With Steak Add to sandwich a tender slice of beef steak which has been grilled or quickly sautéed in butter, with tomato slices. Sautéed mushrooms and onions may also be added.

With Ham and Swiss Cheese Add several slices of boiled ham and Swiss cheese to sandwich before topping with tomato slices. Alfalfa sprouts or endive leaves may also be added. (Turkey or chicken may be included, or used instead of the ham.)

With Cream Cheese and Smoked Salmon Spread a halved bagel, onion roll or a slice of rye bread with a generous amount of cream cheese. Add slices of smoked salmon and freshly ground pepper; top with sliced tomatoes.

With Shrimp or Crab Meat Mix cooked baby shrimp or flaked crab meat with chopped tarragon and chives, a few drops of fresh lemon juice and just enough mayonnaise to bind the mixture. Add to sandwich before topping with tomato slices. (Drained, canned tuna may be used instead of the shrimp or crab meat.)

With Anchovies or Sardines Add canned anchovy fillets or small sardines to sandwich before topping with tomato slices. (French bread is preferable for this sandwich; rings of thinly sliced onion may also be added.)

With Hot Corned Beef or Pastrami Add hot corned beef or pastrami, sauerkraut and Monterey jack cheese to the sandwich; top with tomatillo instead of tomato slices. (Onion rolls or slices of pumpernickel are preferable for this sandwich.)

FRENCH-FRIED TOMATO SANDWICHES

1/2 cup sour cream
1/2 cup grated Gruyère cheese
2 tablespoons each chopped chives and dill
salt and freshly ground pepper to taste
8 thin slices French bread

2 firm, ripe tomatoes, sliced 1/4 inch thick and seeded
2 eggs, beaten
1/4 cup milk
3 tablespoons butter

Mix together the sour cream, cheese, chives, dill, salt and pepper. Butter all of the bread slices on one side and spread the sour cream mixture on the buttered side of 4 slices. Top the sour cream mixture with tomato slices, leaving a 1/4-inch border. Place the remaining buttered bread slices on top, buttered sides in, and press down around edges firmly. Beat together eggs and milk and dip sandwiches in egg-milk mixture. Sauté in butter until golden on both sides, adding more butter as needed. Cut in fourths (triangle shape) and serve hot.
Serves 4

PIZZA

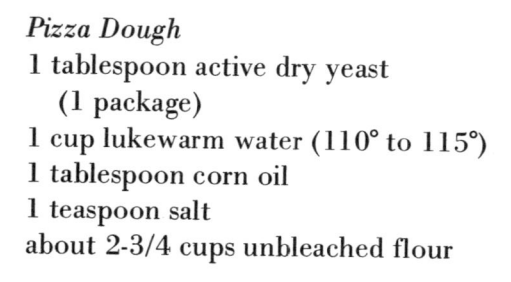

Pizza Dough
1 tablespoon active dry yeast (1 package)
1 cup lukewarm water (110° to 115°)
1 tablespoon corn oil
1 teaspoon salt
about 2-3/4 cups unbleached flour

Tomato Base
3 pounds ripe tomatoes (preferably Italian plum), peeled
1 garlic clove, crushed
2 tablespoons olive oil
1/2 teaspoon sugar

1 teaspoon each dried oregano and basil
1 pound mozzarella or Monterey jack cheese, sliced or grated

Suggested Toppings
Italian salami, very thinly sliced
pepperoni, sliced 1/8-inch thick
sliced olives
cooked Italian sausage, crumbled, or
cooked ground beef, pork or veal
anchovies packed in oil, drained
mushrooms, sliced
sweet red or green pepper, thinly sliced and sautéed
chopped green onions

Soften the yeast in the lukewarm water, then add oil, salt and half the flour, mixing well. Gradually add enough of the remaining flour to form a soft dough. Turn out onto a floured board and knead until smooth and elastic, about 10 minutes. Place in a greased bowl, grease top of dough and cover with a towel. Let rise in a warm place until double in bulk, about 1-1/2 hours.

While dough is rising, prepare the tomato base. Combine the tomatoes, garlic and oil and cook for 10 minutes until tomatoes are tender. Force through a sieve. If tomato pulp is thin, return to pot and cook until slightly thickened. It should not be watery. Stir in sugar.

Turn dough out on a floured board and divide into 2 balls. Roll each into a 13-inch pizza pan or cookie sheet. Spoon half the tomato mixture on each piece of rolled-out dough. Sprinkle with oregano and basil, then top with cheese and any one or more of the suggested toppings. Bake on lowest rack of a preheated 500° oven until browned, about 15 minutes. Cut in wedges and serve immediately.

Makes 2 13-inch pizzas

TOMATO QUICHE

half recipe Pastry Crust, page 78
1/2 cup minced onion
2 garlic cloves, minced
2 tablespoons butter
1-1/2 pounds tomatoes, peeled, seeded and chopped
1 teaspoon sugar
1 whole egg, beaten
3 egg yolks, beaten
3 tablespoons minced parsley
1 tablespoon minced fresh basil
1/2 teaspoon dried oregano
2 tablespoons olive oil
salt and freshly ground pepper to taste
1/2 cup pitted ripe olives, halved
1/3 cup each grated Parmesan and Gruyère cheese

Prepare pastry shell and set aside. Sauté the onion and garlic in butter until onion is translucent. Add tomatoes and sugar and cook over moderate heat until most of the liquid has evaporated. Mixture should be thick. Let cool. Then add egg and egg yolks, parsley, basil, oregano, olive oil, salt and pepper. Pour into pastry shell and garnish top with olives. Sprinkle with cheeses. Bake in a preheated 375° oven for 25 minutes or until set. Let cool 5 minutes before serving.

EGGPLANT PARMIGIANA

3 eggplants (about 1 pound each), peeled and sliced 1/4 inch thick
salt
olive oil
2-1/2 pounds tomatoes, peeled, seeded and chopped
2 tablespoons chopped basil
salt and freshly ground pepper to taste
1 cup grated Parmesan cheese
1 pound mozzarella cheese, sliced or grated

Place the eggplant slices in a colander and lightly salt; let stand for 1 hour. Pat eggplant slices dry with paper toweling and sauté in oil until golden. Drain on paper toweling and set aside. In a saucepan put 2 tablespoons of the oil used to sauté the eggplant. Add tomatoes and basil and cook over medium heat, uncovered, until tomatoes are tender and sauce is thick. Season with salt and pepper. Arrange a layer of eggplant in a shallow baking dish; sprinkle with some of the Parmesan and mozzarella cheese, then some of the tomato sauce. Repeat layers, ending with eggplant, and sprinkle Parmesan over top. Bake in a preheated 350° oven for 45 minutes.

Serves 6 to 8

TOMATO FONDUE

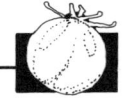

1 garlic clove, bruised
3 tablespoons butter
1 pound tomatoes, peeled, seeded and chopped
1 teaspoon chopped basil

1/4 teaspoon paprika
1 cup dry white wine
2 cups grated cheddar cheese
cubed French bread

Rub the top pan (blazer) of a chafing dish with the garlic clove. Discard. Melt butter in pan and add tomatoes and basil. Simmer 6 to 8 minutes. Add paprika and wine and heat through. Gradually stir in the cheese a little at a time until melted and well blended. Place blazer pan over water pan. Spear bread cubes with a fondue fork and dip in fondue to eat.

Serves 4

TAMALE PIE

1 teaspoon salt
1 cup cornmeal
1 onion, chopped
1 green pepper, chopped
1 or more fresh green chili peppers, chopped
3 tablespoons lard or vegetable shortening
1 pound lean ground beef
1/2 pound chorizo sausage, removed from casing
1 teaspoon chili powder
1 pound tomatoes, peeled and chopped
2 cups freshly cut corn kernels
1/2 pound zucchini, sliced 1/4 inch thick
1 cup pitted ripe olives
2 cups grated cheddar cheese

Bring 3 cups of water to boiling with the salt. Slowly add cornmeal, stirring constantly. Cook on low, stirring often for 10 minutes or until mush is thick. Remove from heat and set aside. Sauté the onion, green pepper and chili pepper in lard until onion is translucent. Add beef, chorizo and chili powder and cook until brown. Add tomatoes, corn and zucchini and cook 10 minutes more. Stir in the olives. Spread the meat mixture in a large shallow greased baking dish and top with cornmeal. Bake in a preheated 375° oven for 20 minutes. Sprinkle cheese on top and bake 10 minutes longer.
Serves 6 to 8

TOMATOES STUFFED WITH TUNA

6 large firm, ripe tomatoes
1 cup fresh bread crumbs
1 onion, finely chopped
2 tablespoons chopped parsley
2 tablespoons chopped ripe olives
2 tablespoons capers
1 7-ounce can tuna with oil
garnish: chopped basil or dill

Prepare tomatoes for stuffing, page 6, lightly salting shells. In a bowl combine the bread crumbs, onion, parsley, olives, capers and tuna and blend well. Fill tomato shells and place in an oiled baking dish. Bake in a preheated 375° oven for 35 minutes. Sprinkle with basil or dill and serve hot.
Serves 6

STUFFED TOMATOES WITH MUSHROOM FILLING

6 large firm, ripe tomatoes
1/2 cup thinly sliced green onions
3/4 pound mushrooms, thinly sliced
4 tablespoons butter
1 tablespoon flour
2 tablespoons fresh lemon juice
1 teaspoon sweet Hungarian paprika
1 teaspoon sugar
3/4 cup heavy cream or evaporated milk
1/4 cup each grated Parmesan and Gruyère cheese, combined

Prepare tomatoes for stuffing, page 6, salting shells and reserving and chopping pulp. Arrange tomatoes cut side up in a buttered shallow baking dish and bake in a preheated 350° oven for 15 minutes. While tomatoes are baking, in a skillet sauté the green onions and mushrooms in butter for 1 minute. Lower heat, add flour and cook and stir 2 minutes. Add reserved tomato pulp, lemon juice, paprika and sugar and sauté 2 minutes. Add cream, blending well, and cook until mixture has thickened. Remove tomatoes from oven and fill with this mixture. Top each tomato with some of the cheese and brown tops under a preheated broiler.
Serves 6

WESTERN BEANS

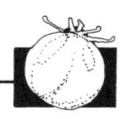

- 1 pound dried pinto, pink or kidney beans, washed and soaked overnight in water to cover
- 1 cracked ham bone, or
- 1 pound ham hocks
- 1 or more dried red chili peppers
- 1 teaspoon dried oregano
- 1 bay leaf
- 2 onions, chopped
- 1 garlic clove, minced
- 4 tablespoons lard or vegetable shortening
- 2 pounds tomatoes, peeled, seeded and chopped
- 1/2 teaspoon ground cumin
- 1-1/2 tablespoons chili powder
- salt and freshly ground pepper to taste

Put the beans and their soaking water, ham bone, chili pepper, oregano and bay leaf in a large pot. Bring to a rapid boil, skim any surface scum and simmer 1-1/2 hours or until beans are just tender. Add boiling water if necessary to prevent beans from drying out. Meanwhile, sauté the onions and garlic in lard or shortening until onions are translucent. Add remaining ingredients to the onions and cook for 10 minutes. Add to the beans and simmer 30 minutes.

Serves 6

BAKED EGGS IN TOMATO SAUCE

1 cup Canned Tomato Sauce, page 84
6 eggs
butter bits
freshly made toast

Put 2 tablespoons tomato sauce into each of 6 ramekins. Drop in a whole egg and cover with 2 teaspoons sauce. Dot with butter. Bake in a preheated 350° oven for 8 to 10 minutes or until eggs are set. Serve at once with toast.
Serves 6

PORK WITH CHILI SALSA

2 pounds pork butt, cut into 1-inch cubes
2 tablespoons flour
2 teaspoons chili powder
3 tablespoons lard
1 pound tomatoes, peeled and chopped
1 onion, chopped
2 garlic cloves, minced
2 or more dried red chili peppers, crumbled
1/2 teaspoon dried oregano
1 teaspoon sugar
1 tablespoon cider vinegar
salt and freshly ground pepper to taste
accompaniments: warm flour or corn tortillas, chopped coriander, shredded lettuce

Lightly dredge the pork pieces with the flour and chili powder. Brown pork in lard, then add all remaining ingredients. Bring to a rapid boil, lower heat to medium, cover and cook for 45 minutes or until pork is tender. Serve in warm tortillas, garnished with coriander and shredded lettuce.
Serves 6

HAM, EGGPLANT, ARTICHOKE AND TOMATO SAUTÉ

2 onions, sliced
1/4 cup olive oil
2 cups diced ham
1 garlic clove, minced
1 green pepper, diced
1 eggplant (about 1 pound), unpeeled and diced
2 pounds tomatoes, peeled and cut in chunks
1 teaspoon salt
1/2 teaspoon dried oregano
2 cups cooked artichoke hearts

Sauté the onions in oil until translucent, then add the ham, garlic, green pepper and eggplant and sauté for 5 minutes. Add the tomatoes, salt and oregano. Cover and cook over medium heat for 10 minutes. Add artichoke hearts and heat through. Serves 6

SHRIMP CREOLE

2 onions, chopped
1 garlic clove, minced
2 green peppers, chopped
1 sweet red pepper, chopped
1 or more dried red chili peppers (optional)
4 tablespoons butter
1-1/2 pounds tomatoes, peeled and chopped
1/2 pound okra, cut up
1/2 teaspoon sugar
1 teaspoon paprika
1-1/2 pounds raw shrimp, shelled and deveined
salt and freshly ground pepper to taste
garnish: chopped parsley
accompaniment: freshly cooked rice

Sauté the onions, garlic, sweet peppers and chili peppers in butter until the onion is translucent. Add tomatoes, okra, sugar and paprika and simmer 20 minutes. Add shrimp and cook 5 minutes. Season with salt and pepper and sprinkle with parsley. Serve over rice.
Serves 4 to 6

CIOPPINO

2 tablespoons each chopped marjoram or oregano, rosemary, sage, thyme and basil
1/2 cup chopped parsley
1 onion, chopped
2 garlic cloves, minced
2 fresh green chili peppers, seeded and chopped
3 cups chopped Swiss chard or spinach
2 Dungeness crabs, cleaned and cracked
1 quart clams or mussels
1 pound raw shrimp in the shell
2 pounds firm fish fillets, cut in 1-1/2-inch pieces
3 pounds tomatoes, peeled and cut up
1/2 cup Canned Tomato Sauce, page 84
2 cups dry red wine
1/2 cup olive oil
salt and freshly ground pepper to taste
accompaniment: crusty French or Italian bread

Mix together the herbs, onion, garlic, peppers and Swiss chard. Put the crabs in the bottom of a large pot. Sprinkle with half of the herb mixture. Next, layer the clams and shrimp and top with remaining herb mixture. Place fish fillets on top and pour in remaining ingredients; cover and simmer 30 minutes. Serve in large soup bowls with crusty French or Italian bread.
Serves 8 to 10

CHICKEN MARENGO

1 3-pound fryer chicken, cut in serving pieces
1/4 cup olive oil
2 garlic cloves, crushed
1-1/2 pounds tomatoes, peeled and quartered
1 cup chicken stock
1/2 cup dry white wine
1 teaspoon dried marjoram
1/4 pound button mushrooms
1 onion, thinly sliced
2 tablespoons butter
1/2 cup pitted black olives
salt and freshly ground pepper to taste
garnish:
    3 tablespoons minced parsley
    juice of 1 lemon

Brown the chicken parts on all sides in the oil. Remove excess oil and set aside. Add the garlic, tomatoes, stock, wine and marjoram. Bring to a rapid boil, then lower heat, cover and simmer 30 minutes. In another pan, sauté the mushrooms and onion in reserved olive oil and the butter for 2 minutes. Add to chicken-tomato mixture along with the olives and continue to cook for 10 minutes longer. Serve garnished with parsley and squeeze juice of lemon over all. Serve immediately.
Serves 4

SAUTEED CHICKEN WITH TOMATOES AND SWEET PEPPERS

1 rosemary sprig
2 garlic cloves, bruised
1/4 cup olive oil
1 3-pound fryer chicken, cut in serving pieces
1 onion, chopped
1 pound tomatoes, peeled, seeded and chopped
1 teaspoon dried oregano
salt and freshly ground pepper to taste
1 pound sweet red peppers

Sauté the rosemary sprig and garlic cloves in the oil until brown; discard sprig and cloves. Sauté the chicken parts in oil until golden on all sides. Add the onion, tomatoes, oregano, salt and pepper. Cover and cook on low heat for 30 minutes. While chicken is cooking, hold the peppers over a flame to blister skin. Remove skin, discard core and slice the peppers into thin strips. Add the peppers to the chicken and continue cooking 15 minutes longer.
Serves 4

VEAL SCALLOPS, SORRENTO STYLE

1 pound milk-fed veal from the leg, sliced thin and pounded to 1/8- to 1/4-inch thickness (12 scallops)
3 tablespoons each olive oil and butter
1 pound tomatoes, peeled, seeded and chopped
2 tablespoons chopped parsley
1 teaspoon chopped oregano
salt and freshly ground pepper to taste
12 paper-thin slices prosciutto
12 thin slices mozzarella cheese
3 tablespoons grated Parmesan cheese

Quickly sauté the veal in half of the butter and oil until lightly browned; do not overcook. Remove veal scallops to a shallow baking dish, placing them in a single layer. In the same pan add remaining oil and butter, tomatoes, parsley, oregano and salt and pepper and cook for 10 minutes over medium heat, or until tomatoes are soft. Place a slice of prosciutto on top of each scallop, then a slice of mozzarella. Spoon the tomato sauce evenly over the scallops. Bake in a preheated 425° oven until the cheese is melted, about 7 minutes. Serve immediately.
Serves 4

LAMB WITH TOMATOES AND SPINACH

2 pounds boneless lamb, cut in 1-inch cubes
3 tablespoons olive oil
1 onion, chopped
1 garlic clove, minced
1/2 teaspoon each dried rosemary and oregano
1 bay leaf
1-1/2 pounds tomatoes, peeled and diced
salt and freshly ground pepper to taste
1 cup water, boiling
2 pounds spinach
1 tablespoon cornstarch, dissolved in 1 tablespoon water

Brown the lamb on all sides in the oil. Add the onion, garlic, rosemary, oregano, bay leaf, tomatoes, salt and pepper and 1 cup boiling water. Cover and simmer 1 hour or until meat is tender. Add the spinach on top, cover and cook 5 to 8 minutes or until spinach is cooked through. Remove the spinach to a warm platter and thicken juices with cornstarch mixture. Ladle lamb over spinach and serve.
Serves 6

CALIFORNIA POT ROAST

2 tablespoons diced salt pork or bacon fat
3 tablespoons olive oil
1 4-pound beef rump roast
2 onions, chopped
1 garlic clove, minced
1 carrot, chopped
1 celery rib with leaves, chopped
1 bay leaf
1 pound tomatoes, peeled, seeded and chopped
1 teaspoon sugar
1 cup red wine
salt and freshly ground pepper to taste

Sauté the salt pork or bacon fat in oil until fat is rendered. Remove bits and brown roast on all sides. Add the onions, garlic, carrot and celery and cook until vegetables begin to brown. Add remaining ingredients, cover and cook on low for 2 hours or until meat is tender. Slice meat and strain sauce to pour over.
Serves 6 to 8

MEATBALLS IN TOMATO SAUCE

Tomato Sauce
1 onion, chopped
3 tablespoons olive oil
1-1/2 pounds tomatoes, peeled, seeded and chopped
1 tablespoon tomato paste
1/2 teaspoon sugar
salt and freshly ground pepper to taste

Meatballs
2 cups soft bread crumbs
1/4 cup milk
1 pound lean ground beef (or half veal)
2 tablespoons minced parsley
1 egg, beaten
3/4 cup grated Parmesan cheese
1 garlic clove, minced
1 teaspoon salt
1/2 teaspoon freshly ground pepper
1/3 cup seedless white raisins
1/3 cup pine nuts or slivered almonds
flour
olive oil

accompaniment: freshly cooked pasta, rice or polenta

Begin tomato sauce by sautéing the onion in olive oil in a saucepan until onion is translucent. Add remaining sauce ingredients, cover and simmer for 40 minutes. While sauce is cooking, combine all the ingredients for meatballs except flour and oil. Blend well and form into balls about 2 inches in diameter. Dredge in flour and fry in oil until brown. Remove with slotted spoon and drain on paper toweling. Drop meatballs into sauce and cook gently 10 minutes. Serve with accompaniment.
Serves 4

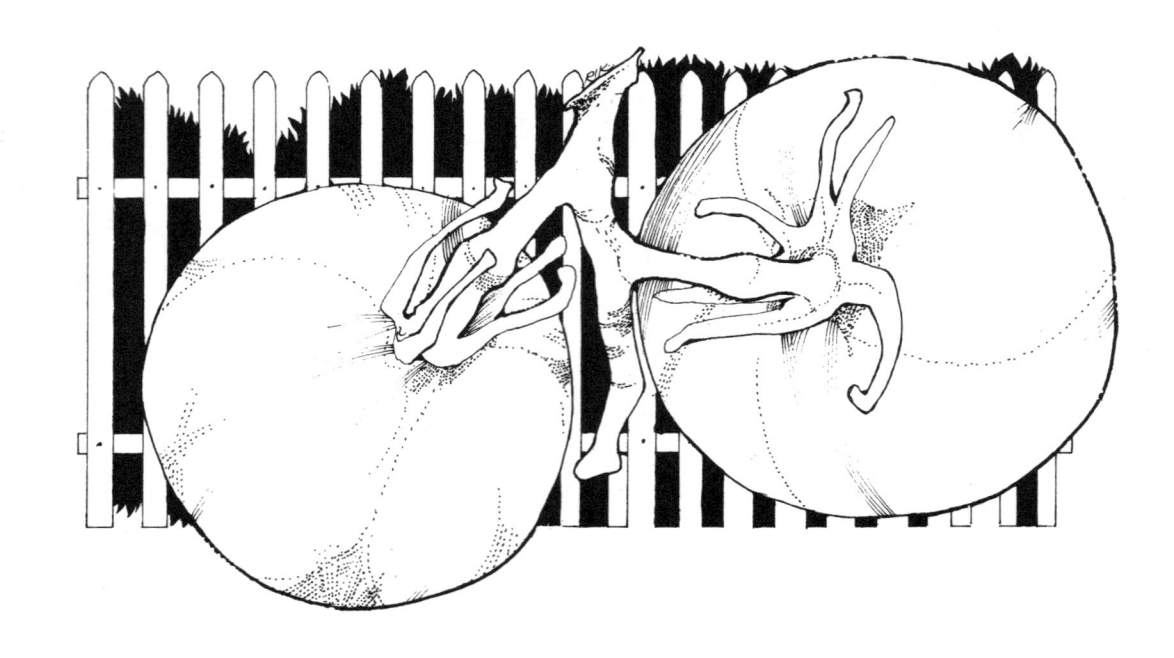

sauces

BASIC TOMATO SAUCE

1 carrot, cut up
2 garlic cloves
1 onion, cut up
3 pounds tomatoes, peeled, seeded and cut up (preferably Italian plum)
3 parsley sprigs
1 thyme sprig
1 rosemary sprig
1 bay leaf
1/2 cup olive oil
salt and freshly ground pepper to taste

In several batches, put the vegetables and herbs in a blender to chop coarsely. Put vegetable-herb mixture in a saucepan with olive oil, bring to a rapid boil, then lower heat and simmer, covered, for 30 minutes. Press through a sieve and return to saucepan. Add salt and pepper and simmer the sauce, uncovered, stirring often for 40 minutes or until thick. May be stored in refrigerator for 1 week, or frozen.
Makes 3 to 4 cups

MARINARA SAUCE

2 garlic cloves, minced
1 medium onion, chopped
1/2 pound mushrooms, sliced, or
1/2 cup dried Italian mushrooms, soaked in warm water to soften and minced
1/2 cup olive oil
2 pounds tomatoes, peeled, seeded and chopped (preferably Italian plum)
1 cup Canned Tomato Paste, page 84

3/4 cup water
3/4 cup dry red wine
1 teaspoon salt
1 teaspoon minced oregano
1 tablespoon minced basil
1/2 teaspoon rosemary
1/2 teaspoon freshly ground pepper
1/4 teaspoon ground cinnamon
1/2 cup chopped Italian parsley

Sauté garlic, onion and mushrooms in oil until onion is translucent. Add all remaining ingredients and simmer, uncovered, for 1-1/2 hours. Serve over any pasta or ravioli.
Makes 4 to 6 cups

Variation with Meat During the last 15 minutes of cooking, add 1 pound ground beef or veal that has been sautéed in 2 tablespoons olive oil.

QUICK TOMATO SAUCE FOR PASTA

3/4 cup olive oil
3 garlic cloves, crushed
2 pounds tomatoes, peeled, seeded and chopped (preferably Italian plum)
1/2 cup dry white wine
3 tablespoons chopped basil

2 tablespoons chopped parsley
salt and freshly ground pepper to taste
freshly cooked pasta (12 ounces dry or 1 pound fresh)
freshly grated Parmesan cheese

Heat the oil and sauté garlic until brown. Discard garlic and add tomatoes, wine, basil, parsley, salt and pepper. Cook for 20 minutes over medium heat. Toss with pasta and sprinkle with cheese. Serve immediately.
Serves 4

With Anchovies Reduce oil to 1/2 cup. Proceed as above, adding 1 2-ounce tin anchovy fillets, drained, after discarding garlic. Mash anchovies in the oil until they form a paste. Add remaining ingredients, omitting basil, and proceed as directed.

With Squid Proceed as above. Add 2 pounds squid, cleaned and cut into rings, with tomatoes.

With Shrimp Proceed as above, reducing basil to 1 tablespoon and adding 1 pound raw shrimp, shelled and deveined, to the sauce the last 5 minutes of cooking.

With Clams Mince the garlic and sauté with 1 onion, chopped. Do not discard garlic. Proceed as directed above, omitting basil and adding 2 cups canned and drained baby clams to the sauce at the end of cooking; heat through and serve.

With Zucchini or Eggplant Proceed as above, reducing basil to 1 tablespoon and adding 1 teaspoon dried oregano, 1 pound diced zucchini or eggplant and 1 green pepper, chopped, with the tomatoes.

With Beef or Veal Reduce oil to 1/2 cup. Mince garlic and sauté it with 1/2 pound ground beef or veal and 1 small onion, chopped; do not discard garlic. Proceed as directed above.

With Mushrooms Sauté 1/2 pound mushrooms, chopped, in the oil after discarding garlic. Proceed as directed above, reducing basil to 1 tablespoon and adding 1/2 teaspoon dried oregano and 1/4 teaspoon ground cinnamon with the tomatoes.

With Bacon Reduce oil to 2 tablespoons and sauté 6 slices bacon, diced, and 1 onion, chopped, in the oil after discarding garlic. When bacon is cooked proceed as directed above, omitting basil and adding 1 teaspoon dried marjoram with the tomatoes.

TOMATO SAUCE PROVENÇALE

1 onion, minced
2 garlic cloves, minced
1 green pepper, chopped
1/4 cup olive oil
3/4 cup dry white wine
2 pounds tomatoes, peeled, seeded and chopped
1/3 cup minced parsley
2 teaspoons minced basil
1/2 teaspoon sugar
1/4 teaspoon crumbled saffron threads
salt and freshly ground pepper to taste

Combine all ingredients except salt and pepper in a saucepan and bring to a boil. Lower heat and simmer, uncovered, for 20 minutes or until thick and the liquid has been reduced by half. Season with salt and pepper. Good over sautéed or broiled chicken or seafood, or cooked vegetables.

Makes 2 to 3 cups

CHICKEN LIVER-MUSHROOM TOMATO SAUCE

1/2 pound chicken livers, halved
6 tablespoons butter
1 onion, minced
1 garlic clove, minced
1/4 pound mushrooms, thinly sliced
1-1/2 pounds tomatoes, peeled, seeded and chopped (preferably Italian plum)
1/2 cup dry red wine
2 tablespoons Canned Tomato Paste, page 84
1/2 teaspoon dried thyme
1/2 teaspoon dried tarragon
1/4 cup minced parsley

In a skillet sauté the chicken livers in 2 tablespoons of the butter until brown; remove from pan and set aside. Return skillet to heat, add remaining butter and sauté onion, garlic and mushrooms for 5 minutes. Add the tomatoes, wine, tomato paste, thyme and tarragon and simmer for 15 minutes or until thick. Add the livers, heat through and stir in parsley. Serve over cooked pasta, polenta or rice.
Makes about 5 cups

CREOLE SAUCE

1 onion, minced
1 garlic clove, minced
1/2 cup chopped green pepper
2 tablespoons butter
1 bay leaf, crumbled
2 pounds tomatoes, peeled, seeded and chopped
1 teaspoon salt
1/4 teaspoon freshly ground pepper
1/4 teaspoon dried thyme

Sauté the onion, garlic and green pepper in butter until onion is translucent. Add remaining ingredients and cook over moderate heat until mixture is thick and has been reduced to one-half. Good with seafood, poultry, meat or eggs.
Makes about 2 cups

TOMATO ORANGE SAUCE

1/3 cup sliced celery
2 tablespoons each butter and flour
1 cup peeled, seeded and chopped tomatoes

1/2 cup each orange juice and water
1/2 teaspoon each sugar and salt
1 teaspoon grated orange peel
1/8 teaspoon freshly grated nutmeg

Sauté celery in butter until tender. Stir in flour and cook 2 minutes. Add tomatoes, orange juice and water and cook until thickened. Add remaining ingredients and heat through. Good on vegetables, poultry or fish.
Makes about 2 cups

TOMATO-DILL SOUR CREAM SAUCE

1 onion, chopped
2 carrots, chopped
4 tablespoons butter
2 pounds tomatoes, peeled, seeded and chopped
1 tablespoon flour

3/4 cup chicken stock
1 teaspoon sugar
1/2 cup sour cream
3 tablespoons chopped dill
salt and freshly ground pepper to taste

Sauté the onion and carrots in 2 tablespoons of the butter for 5 minutes. Add the tomatoes and simmer 15 minutes, uncovered, until thick and most of the liquid has evaporated. Let mixture cool a bit, then blend in a blender. In a skillet melt remaining butter, add flour and cook on low 2 minutes, stirring constantly. Add tomato mixture, stock and sugar and cook 10 minutes until thick and smooth. Remove from heat and stir in sour cream, dill, salt and pepper. Serve this sauce hot or cold on vegetables, salads, poached salmon, shrimp, turkey or chicken.
Makes about 3 cups

TOMATO BARBECUE SAUCE

1-1/2 pounds tomatoes, peeled and chopped
2 tablespoons olive oil
1 onion, chopped
1 garlic clove, chopped
1 or 2 slices ginger root, minced (optional)

1 tablespoon brown sugar
1 cup Canned Tomato Sauce, page 84
1/4 cup each Worcestershire sauce and cider vinegar
1/2 teaspoon dry mustard
1 teaspoon salt
1/2 teaspoon freshly ground pepper

Combine all ingredients and simmer for 30 minutes. Use to marinate and baste chicken, ribs, steaks, chops. Store in a covered jar in the refrigerator.
Makes about 2 cups

WESTERN BARBECUE SAUCE

1 cup Canned Tomato Sauce, page 84
2 cups water
2 tablespoons olive oil
2 garlic cloves, minced
1 onion, minced
1 or more dried red chili peppers
1 carrot, grated
2 bay leaves

4 whole cloves
1 teaspoon dried oregano
1/2 teaspoon ground cumin
2 tablespoons cider vinegar
1 tablespoon sugar
1 teaspoon freshly ground pepper
salt to taste

Combine all ingredients and simmer for 30 minutes. Let stand until cool, then strain the sauce. This sauce may be used as a marinade or for basting any meat or poultry.
Makes about 2 cups

MEXICAN TOMATO SALSA

1 pound ripe, juicy tomatoes, peeled and chopped
1 or more fresh green chili peppers, minced
1 small onion, minced
1 garlic clove, minced

2 tablespoons each cider vinegar and olive oil
1 tablespoon minced coriander leaves
1/2 teaspoon dried oregano
salt to taste

Combine all ingredients and let stand 2 hours at room temperature to blend flavors. Store in refrigerator. Use on tacos, tostadas, barbecued meats, Mexican beans, meat dishes or as a spicy dressing for green salads.
Makes about 2 cups

GREEN TACO SAUCE

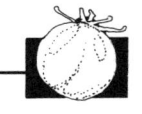

1 pound tomatillos, husks removed, simmered in water to cover 10 minutes or until soft and drained
1/4 cup coriander leaves
1 tablespoon fresh lime juice

1 garlic clove
3 tablespoons chopped onion
4 tablespoons canned diced green chili peppers

Place all the ingredients in a blender and blend until smooth. Store covered in refrigerator.
Makes about 2 cups

sweets & pastries

CANDIED PLUM TOMATOES

2 pounds ripe Italian plum tomatoes
2 cups firmly packed light brown sugar
1/8 teaspoon ground cloves

1/4 teaspoon ground cinnamon
powdered sugar

Peel and core tomatoes; let stand to drain well. Melt the sugar over low heat in a heavy saucepan. Add cloves, cinnamon and tomatoes and cook on low until tomatoes are clear and candied. Put candied tomatoes in a shallow pan and place in a preheated 200° oven until tomatoes are dried, about 1-1/2 hours. Tomatoes should have texture of dried apricots. Remove from pan and roll each candied tomato in powdered sugar. Store in airtight container.
Makes 1 pound

TOMATO SPICE CAKE

6 tablespoons butter
1/2 cup each white and firmly packed brown sugar
2 cups unbleached flour
1 teaspoon baking soda
1/2 teaspoon salt
1 teaspoon ground cinnamon
1/2 teaspoon ground nutmeg
1/4 teaspoon ground cloves
1 cup plus 2 tablespoons Canned Tomato Juice, page 87
1 cup chopped walnuts or pecans
powdered sugar or white icing of choice

Cream the butter with the sugars. Sift together flour, soda, salt and spices and add to the sugar-butter mixture alternately with the juice. Fold in nuts and pour batter into a well-greased and lightly floured 9- by 5-inch loaf pan. Bake in a preheated 350° oven for about 1 hour or until a toothpick inserted in the center comes out clean. Dust top with powdered sugar when cool or frost with a white icing.

MINCEMEAT DROP COOKIES

1/4 pound butter, at room temperature
1 cup firmly packed brown sugar
2 eggs, beaten
1-1/2 cups Green Tomato Mincemeat, page 92
2-1/2 cups unbleached flour
1 teaspoon baking soda
1 cup chopped walnuts or pecans

Cream the butter and sugar until light and fluffy. Beat in the eggs one at a time, then beat in mincemeat. Sift together the flour and baking soda and stir into batter, blending well. Drop by teaspoonfuls onto a greased cookie sheet and bake in a preheated 400° oven for 10 minutes.
Makes 5 dozen cookies

VERMONT GREEN TOMATO PIE

Pastry Crust, following
2 pounds green tomatoes, thinly sliced
1 cup sugar
1/2 teaspoon each ground cinnamon and allspice
1 tablespoon cornstarch
1/4 cup cider vinegar
2 tablespoons butter

Prepare pastry dough, roll out half as directed in recipe and fit into a 9-inch pie pan. Place a layer of tomato slices on the pastry shell. Combine the sugar, spices and cornstarch and sprinkle some of this mixture on top of the tomatoes; repeat layers until tomatoes and spice mixture are used up. Sprinkle vinegar over top and dot with butter. Roll out remaining pastry, fit top of pie, flute edges together and make a vent slit. Bake in a preheated 425° oven for 20 minutes. Reduce oven to 375° and continue baking for 40 minutes. Serve lukewarm with a slice of cheddar cheese, if desired.

PASTRY CRUST

2 cups unbleached flour
1 teaspoon salt
3/4 cup vegetable shortening or lard, or half butter and half shortening
5 to 6 tablespoons cold water

Put the flour and salt in a bowl and cut in shortening with a pastry blender or two knives until mixture is crumbly. Sprinkle water over mixture, mixing lightly with a fork. Form into a ball with hands and divide in half. Roll out each half on a lightly floured board to a thickness of 1/8 inch. Fit into pie plates or on top of pie according to recipe. For one crust, halve the above ingredients and proceed as directed.

Makes two 8- or 9-inch crusts

TOMATO BISCUITS

2 cups unbleached flour
2 teaspoons baking powder
1/4 teaspoon baking soda

1 teaspoon salt
1/3 cup corn oil
2/3 cup Canned Tomato Juice, page 87

Sift together the flour, baking powder, soda and salt into a mixing bowl. Pour in the oil and tomato juice and stir with a fork until dough is well mixed. Place dough on a lightly floured board and knead about 10 times. Roll out to 1/2-inch thickness and cut into rounds with a biscuit cutter. Place on an ungreased baking sheet and bake in a preheated 475° oven for 10 to 12 minutes or until golden. Serve immediately with butter and honey, jelly or jam.

Makes 16 2-inch biscuits

Tomato-Herb Biscuits Add 1 tablespoon chopped chives and 1 teaspoon chopped basil or dill to the tomato juice.

Tomato-Cheese Biscuits Add 1/2 cup grated cheddar cheese to the flour mixture.

preserving

GENERAL PROCEDURES FOR CANNING

All equipment and utensils (except canning lids with sealing compound attached, see below) must be first washed in hot soapy water, then sterilized by boiling in water to cover for 20 minutes and left in the hot water until used. You will need: nonmetal cups; stainless steel or enamel spoons; a wide-mouthed funnel; tongs or a jar lifter; and tempered glass canning jars with lids.

There are three types of canning jars commonly used. Ones with metal lids with an attached rubber-like sealing compound and separate ring bands require that new lids be used each time. Only the ring bands, not the lids, should be sterilized. This jar automatically creates a vacuum as it is processed and the ring band does not need to be tightened after processing. Glass lids with wire bail closures and separate rubber rings, and zinc lids with separate rubber rings both require new rubber rings with each use. During processing glass lids are held in place over the rubber rings by the wire bail on top of the lid; sealing is completed after processing by pushing the short wire down. Zinc lids are held in place over rubber rings during processing by screwing lids on firmly, then turning back 1/4 inch. The seal is completed by screwing lid on tightly after processing. Whichever type of canning jar you choose, remember to read the manufacturer's instructions for any special information.

With all three types of canning jars, it is advised to check closure before using by filling jars with water, closing and inverting. If water leaks out, the jars are not safe to use. Be sure with all types of canning jars to wipe rim of jar carefully with a *clean* cloth after filling so that no food will remain to affect closure. Remove from processor with tongs or jar lifter attached to jars, not to lids. Cool processed jars on a rack or folded towel, away from any drafts, spacing them several inches apart. It is important to leave all jars undisturbed for at least *12 hours* after processing. Then check lids: Those properly sealed will "ting" when tapped with a metal spoon; those not properly sealed will make a hollow sound and must be used immediately or destroyed.

If tomatoes are cooked before they are put into sterilized canning jars, the kettle in which they are initially cooked, as well as the spoons used to stir them, must be made of stainless steel or enamel which is free of any chips or cracks. Never use copper, brass or iron kettles or utensils.

PROCESSING IN A BOILING WATER BATH

If the recipe calls for processing the canning jars in a boiling water bath, you will need a kettle with a rack and a tight-fitting lid, large enough to hold jars with 2 inches of space between them and 2 inches of water to cover. For the cold pack method the jars are placed in a kettle half-filled with hot water; for the hot pack method the jars are placed in a kettle half-filled with boiling water. Water is then added to cover the jars. It should always be directed down the sides of the kettle and never directly on jar tops. Bring water to a rolling boil, then cover kettle and process the jars for the length of time specified in the recipe you are using. If necessary, add boiling water during processing to maintain the water level.

PROCESSING IN A PRESSURE CANNER

Tomatoes have been traditionally thought of as an acid food. However, the proliferation of new varieties has raised the possibility of varying levels of acidity. To eliminate any doubt as to the safety of home-canned tomatoes, add 1 tablespoon lemon juice per pint of tomatoes to be canned. Tomatoes canned alone, with no other ingredient except the added lemon juice (and salt, if desired), may be processed in a boiling water bath. Tomatoes mixed with vegetables and vinegar with spices, are also processed in a boiling water bath. However, tomatoes mixed with vegetables or meat, as in stewed tomatoes or sauces, are subject to possible bacterial interaction, and *must* be processed *only* in a pressure canner, which is especially designed for high-temperature processing.

Follow the manufacturer's instructions carefully when using a pressure canner. Be sure the canner is clean and in good condition. Before closing the petcock or steam vent it is important to exhaust the air from the canner for at least 10 minutes. Pressure requirements for canning are figured at sea level; more pressure is needed at higher altitudes. One-half pound must be added to the pressure gauge for each additional 1000 feet.

CANNING TOMATOES

Choose only perfect, firm, full flavored and freshly picked tomatoes. Be sure the tomatoes are not overripe. It is best to can them the day they are picked. Sort tomatoes by size for uniformity in heat distribution during processing. Have all your canning equipment ready; try to complete the canning process without waiting between steps, as tomatoes should not be exposed to the air during canning any longer than necessary. First peel the tomatoes (see Peeling Tomatoes, page 6), then cut out the stem core. Tomatoes may be left whole, halved or quartered. The cold pack method is best for canning whole tomatoes.

Cold Pack Method Peel and core tomatoes and pack raw into hot sterilized canning jars, pressing tomatoes down gently with a spatula to eliminate air pockets, and allow 1/2-inch headspace. Add lemon juice (1 tablespoon per pint) to each jar. If you wish, add 1/2 teaspoon salt per pint. Wipe rims, seal jars, place in a kettle which has been half-filled with hot water, spacing jars 2 inches apart, and add hot water to cover jars by 2 inches. Bring to a rolling boil, then cover and process 40 minutes for pints and 50 minutes for quarts.

Hot Pack Method Peel and core tomatoes and put in a stainless steel or enamel kettle. Bring to a boil, then pack into hot sterilized jars, leaving 1/2-inch headspace. Add 1 tablespoon of lemon juice per pint, and if desired, 1/2 teaspoon salt per pint of tomatoes. Run a spatula along the inside of each jar to remove any air pockets. Wipe rims, seal jars, place in a kettle which has been half-filled with boiling water, spacing jars 2 inches apart, and add boiling water to cover jars by 2 inches. Bring to a rolling boil, then cover and process 35 minutes for pints and 45 minutes for quarts.

CANNING TOMATO PICKLES AND RELISHES AND JAMS AND JELLIES

When canning tomato pickles and relishes, follow the rules for sterilizing equipment, and filling, sealing, processing in a boiling water bath and storing jars. You may use either cider vinegar or distilled vinegar of 4 to 6 percent acidity. Recipes for pickled cucumbers usually specify pickling salt and soft or distilled water to obtain a bright pickle with a clear liquid; however, regular table salt and tap water may be used for all the tomato pickling recipes in this book unless otherwise specified. For all pickles and relishes, be sure to leave 1/2-inch headspace at the top of each jar. Pickles and relishes, like other canned foods, should be stored in a cool, dark place. Their flavor is improved if they are stored for at least 1 month before eating.

Tomato jams and jellies do not need to be processed in a boiling water bath. Be sure, however, to use fresh, perfect fruit and sterilized equipment and store properly. Leave a 1/8-inch headspace for all jams and jellies.

CANNED TOMATO SAUCE

3 sweet red or green peppers
3 carrots, cut up
6 garlic cloves
3 onions, cut up
10 pounds tomatoes, peeled, seeded and cut up

4 parsley sprigs
2 thyme sprigs
2 rosemary sprigs
3 bay leaves
1 tablespoon salt

In several batches, put the vegetables in a blender to chop coarsely. Put vegetables and remaining ingredients into a kettle; bring to a rapid boil, then lower heat and simmer, covered, for 30 minutes. Press through a sieve, return to kettle and continue to simmer the sauce, uncovered, stirring often for 40 minutes or until thick. Pour into hot sterilized pint jars, leaving 1/2-inch headspace, and seal immediately. Process in a pressure canner for 20 minutes under 10 pounds pressure.
Makes about 5 pints

Note Basic Tomato Sauce, page 68, or a commercially canned tomato sauce may be substituted in recipes calling for Canned Tomato Sauce.

CANNED TOMATO PASTE

25 pounds tomatoes, peeled and cut up (preferably Italian plum)
1 tablespoon salt

approximately 1/4 cup fresh lemon juice

Combine tomatoes and salt in a large kettle and simmer 1 hour. Put through a sieve. Return to the kettle and continue to cook over low heat about 2-1/2 hours until very thick and a spoonful of the mixture forms a mound when spooned onto a plate. Pour into hot sterilized half-pint jars, leaving 1/2-inch headspace. Mix 1 teaspoon lemon juice into each jar, seal and process in a boiling water bath for 15 minutes. Makes about 10 half pints

Note Dried Tomato Paste, following, Tomato Paste Balls, page 86 or commercially canned tomato paste may be substituted in recipes calling for Canned Tomato Paste.

DRIED TOMATO PASTE

4 quarts peeled and quartered ripe tomatoes (about 8 pounds—preferably Italian plum)
2 tablespoons chopped basil
2 teaspoons salt
1/2 cup chopped celery

1 cup sliced carrots
1 onion, sliced
1 2-inch cinnamon stick
1/2 teaspoon peppercorns
1/2 teaspoon whole cloves

Combine all ingredients in a large enamel pot and simmer until soft, about 1 hour. Put through a sieve. Return mixture to pot, place pot over an asbestos pad and simmer uncovered until very thick, about 3 hours. Spread to 1/2-inch thickness on platters which have been rinsed with cold water. Cover with a fine-mesh screen and set in the sun to dry. This may take up to 1 week, depending on temperature and humidity levels. The tomato paste should be very dry and brittle before storing. Break up into pieces and store in containers in a dry, cool place. Dissolve in a small amount of boiling water and use to flavor soups, sauces and stews.

TOMATO PASTE BALLS

Proceed as for Dried Tomato Paste, preceding. When partially dry (after a few days of drying), enough to form into a stiff paste, form into balls, moisten each ball with olive oil and place the balls in a stone or glass jar. Place a clean cloth dipped in olive oil in the top of the jar to prevent the balls from drying out. Cover jar tightly. Will keep indefinitely. Use by dissolving a small amount in boiling water.

TOMATO CATSUP

10 pounds ripe tomatoes, peeled, seeded and chopped
2 sweet red peppers, chopped
2 onions, chopped
1/4 teaspoon cayenne pepper
1 tablespoon salt

1 cup sugar
1 cup cider vinegar
spice bag (tied in cheesecloth) of:
    2 cinnamon sticks
    1 teaspoon each whole cloves, allspice, celery seed and mustard seed

In a large kettle combine the tomatoes, peppers and onions. Bring to a slow boil and cook 20 minutes or until vegetables are soft. Cool slightly and purée in a blender or force through a food mill. Return puréed mixture to the kettle, add cayenne, salt and sugar and cook on low until mixture is reduced by half, about 1-1/2 to 2 hours. Add vinegar and spice bag and continue cooking on low until very thick or of desired consistency. Remove spice bag and pour into hot sterilized jars. Seal immediately and process for 15 minutes in a boiling water bath.
Makes about 4 pints

Garlicky Tomato Catsup Omit spice bag and add 1/2 cup minced garlic and 1/4 cup minced ginger root with the sugar.

ENGLISH CHILI SAUCE

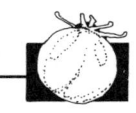

4 pounds ripe tomatoes, peeled and chopped
4 pounds tart apples, peeled and chopped
4 pounds onions, chopped
4 green peppers, chopped
4 sweet red peppers, chopped
1 cup raisins, chopped
2 cups cider vinegar
1 cup sugar
2 tablespoons salt
1 teaspoon ground cinnamon
1/2 teaspoon each ground cloves, ginger and nutmeg

Combine all the ingredients and cook slowly until thick, about 1 hour, stirring frequently. Pour into hot sterilized jars and seal immediately. Process for 15 minutes in a boiling water bath.
Makes about 5 pints

CANNED TOMATO JUICE

Peel, core and chop tomatoes. Simmer in a stainless steel or enamel kettle for 30 minutes, stirring occasionally. Put through a sieve and discard pulp and seeds. Return juice to the kettle and to each quart add 1 teaspoon salt, if desired, and 2 tablespoons of fresh lemon juice. Bring to a boil and pour boiling hot into hot sterilized jars, leaving 1/2-inch headspace. Seal and process for 15 minutes in a boiling water bath, for both pints and quarts.

Note To freeze tomato juice, prepare as above. After cooking juice down, let cool, pour into plastic or glass containers with tight-fitting lids and place in the freezer. See also recipe for Quick Tomato Juice, page 30.

KOSHER DILL GREEN TOMATOES

Brine
1 cup kosher salt or sea salt
3 quarts water
1 quart cider vinegar

8 garlic cloves
8 dried red chili peppers
peppercorns
8 dill sprigs
8 fresh grape leaves
10 pounds small whole green tomatoes, stemmed

Combine the brine ingredients and bring to a boil. Keep hot. Have ready 8 hot sterilized quart jars. To each jar add 1 garlic clove, 1 dried red chili pepper, 3 or 4 peppercorns, 1 sprig of fresh dill and 1 grape leaf and fill with tomatoes. Insert a stainless-steel knife blade in the jar and pour in hot brine to fill the jar, leaving 1/4-inch headspace. Remove knife, seal, and process for 15 minutes in a boiling water bath.
Makes 8 quarts

PICCALILLI

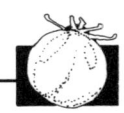

5 pounds green tomatoes
2 sweet red peppers
2 green peppers
2 pounds onions
1 small head cabbage
1/2 cup salt
spice bag (tied in cheesecloth):
    1 tablespoon each celery seed and mustard seed

    1 teaspoon each whole cloves and allspice
1 2-inch cinnamon stick
1 quart cider vinegar
3 cups brown sugar
1 tablespoon prepared horseradish

Chop all the vegetables coarsely and mix with salt. Let stand overnight. Drain thoroughly. In a large kettle combine the spice bag, vinegar, sugar and horseradish. Bring to a boil, add the vegetables and cook on low 20 minutes. Ladle into hot sterilized jars and seal. Process for 15 minutes in a boiling water bath.
Makes about 4 to 6 pints

OLD-FASHIONED GREEN TOMATO PICKLE

4 quarts thinly sliced green tomatoes
4 onions, thinly sliced
1/2 cup kosher salt or sea salt
2 cups brown sugar
2 cups cider vinegar

1/2 teaspoon ground allspice
1 tablespoon celery seed
4 tablespoons mustard seed
1 tablespoon ground turmeric

Combine the tomatoes and onions and sprinkle with salt. Let stand 4 hours. Drain thoroughly. In a kettle combine the remaining ingredients and bring to a boil. Add vegetables and return to a boil; reduce heat and cook on low for 5 minutes. Ladle into hot sterilized jars and seal immediately. Process for 15 minutes in a boiling water bath.
Makes about 4 quarts

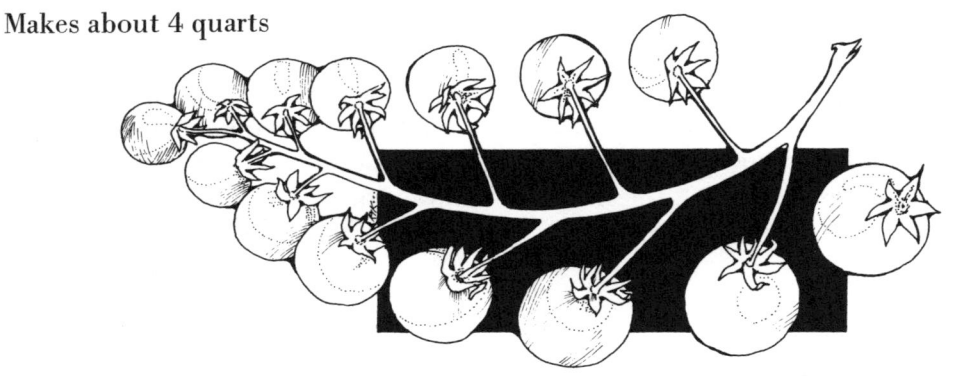

TOMATO JELLY

Slice tomatoes which are almost ripe. Cook slowly over low heat until tender. Using a jelly bag or a piece of muslin which has been immersed in water and wrung out, then used to line a strainer or colander, allow the tomatoes to drain into a container. If you wish a clear jelly, do not squeeze the bag. For every quart of tomato liquid obtained add the rind of 1 lemon and boil for 20 minutes. Discard rind. Measure 1 cup sugar for every cup of liquid. Place sugar in the oven until it is warm, then add to the liquid and cook to the jelly stage, 220° to 222° on a candy thermometer, or when liquid falls in a sheet from the side of a spoon. Fill hot sterilized jars and seal.

TOMATO MARMALADE

5 pounds tomatoes, seeded and chopped
2 large oranges, peeled and peel cut in thin julienne
2 lemons, peeled and peel cut in thin julienne
2 cinnamon sticks
1 teaspoon whole cloves
8 cups sugar

Place the tomatoes in a colander to drain excess juice. Discard pith and seeds from oranges and lemons and chop the fruit. Tie cinnamon sticks and cloves in a piece of cheesecloth. Put the drained tomatoes, oranges and lemons, their rind, the spice bag and sugar in a large saucepan or kettle and cook over low heat until sugar is dissolved, stirring occasionally, for about 25 minutes. Increase heat to high and boil the mixture for 30 minutes or until thick. Discard spice bag, ladle into hot sterilized jars and seal.
Makes about 3 pints

TOMATO BUTTER

- 2 pounds red tomatoes, peeled and chopped
- 3 pounds green tomatoes, peeled and chopped
- 2 lemons, halved and thinly sliced (including peel), seeds removed
- 3 cups sugar
- 1/2 teaspoon ground cloves
- 2 tablespoons minced fresh ginger root or crystallized ginger
- 2 tablespoons chopped candied orange peel

In a large kettle combine all ingredients, bring to a slow boil and cook over moderate heat until thick, about 45 minutes. Ladle into hot sterilized jars and seal.
Makes about 3 pints

TOMATO CONSERVE

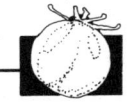

- 4 pounds ripe tomatoes, peeled and seeded
- 4 cups sugar
- 1 cup raisins
- 1 orange, seeded and thinly sliced (including peel)
- 1 lemon, seeded and thinly sliced (including peel)
- 1 cup chopped walnuts

Combine the tomatoes and sugar and cook on low for 1-1/2 hours or until thick. Add raisins, orange and lemon and cook 30 minutes more. Add nuts and cook 5 minutes. Pour into hot sterilized jars and seal.
Makes about 3 pints

MIXED FRUIT CHUTNEY

5 pounds tomatoes, peeled, seeded and chopped
2 pounds peaches, peeled and chopped
2 pounds pears, peeled and chopped
3 onions, chopped
2 green peppers, chopped
1 or more dried red chili peppers, crushed
2 tablespoons chopped fresh ginger root or crystallized ginger
1-1/2 tablespoons salt
3 cups sugar
3 cups cider vinegar
1/4 cup mixed pickling spices tied in cheesecloth

In a large enamel kettle combine all the ingredients and bring to a boil, stirring constantly. Reduce heat to low and simmer, uncovered, for 2-1/2 hours or until thick, stirring occasionally. Remove spice bag, ladle into hot sterilized jars and seal. Process for 15 minutes in a boiling water bath.
Makes about 5 pints

GREEN TOMATO MINCEMEAT

4 quarts chopped green tomatoes
4 quarts peeled and chopped tart apples
8 cups sugar
1 pound each raisins and currants, chopped
2 oranges, seeded and chopped (including peel)
1 lemon, seeded and chopped (including peel)
1 tablespoon ground cinnamon
2 tablespoons ground allspice
1 teaspoon ground cloves
2 tablespoons salt
2 cups apple cider
2 cups apple brandy

Combine all ingredients. Bring to a slow boil and simmer until thick, about 2 hours, stirring occasionally. Pour into hot sterilized jars and seal. Process for 15 minutes in a boiling water bath. This recipe is excellent for pie: Try using half mincemeat and half sliced apples.

Makes about 4 quarts

FREEZING TOMATOES

Tomatoes should be prepared for freezing as soon as possible after picking for best flavor and to retain the most vitamins. Choose only perfect, firm, ripe full-flavored tomatoes. Wipe them clean and put into airtight plastic bags or rigid containers, leaving 1/2-inch headspace. Date the containers and freeze quickly. Use within a couple of months for best flavor and nutritive value. After freezing the skins will slip off easily when placed under cold running water. Frozen tomatoes may be used in any recipe requiring cooking or where the texture of a fresh tomato is not necessary. I doubt that you would like the texture of frozen, thawed tomatoes in salads.

Tomato sauces, soups, stews, etc. all freeze very well. Simply prepare your favorite recipe, put into airtight containers, glass or plastic, label them and their date and freeze. Use within a year's time; preferably sooner.

TOMATO LOTION FOR THE SKIN

Combine 1/3 cup freshly squeezed, strained tomato juice, 1/3 cup fresh lime or lemon juice and 1/3 cup rosewater (or 1/3 cup olive oil or glycerine for an oil-based lotion). Place mixture in a clean jar, screw on cover and shake well. Store in the refrigerator. Use on your face and hands as a refreshing tonic for the skin.

Makes 1 cup

index to recipes

Aspics
  Quick Tomato Aspic, 43
  Tomato Aspic, 42-43
  Tomato Aspic Freeze, 43

Bacon, Tomato and Green Bean Sauté, 47
Baked Cheddar Tomatoes, 46
Baked Eggs in Tomato Sauce, 61
Baked Sliced Tomatoes, 45
Barbecue Sauce, Tomato, 74
Barbecue Sauce, Western, 74
Basic Tomato Sauce, 68-69
Beans, Western, 60
Beef and Veal
  California Pot Roast, 66-67
  Meatballs in Tomato Sauce, 67
  Tamale Pie, 58
  Veal Scallops, Sorrento Style, 65
Beverages
  Bloody Mary with Variations, 31
  Canned Tomato Juice, 87
  Quick Tomato Juice, 30
  Tomato Juice Beverages, 30-31
Biscuits, Tomato, with Variations, 79
Bloody Mary with Variations, 31
Bouillon, Tomato, 25
Broiled Tomato Halves, 46
Butter, Tomato, 91

Cake, Tomato Spice, 77
Candied Plum Tomatoes, 76
Canned Tomato Juice, 87
Canned Tomato Paste, 84-85
Canned Tomato Sauce, 84
Canning, General Procedures for, 80-84
Catsup, Tomato, 86
Chicken
  Chicken Marengo, 64
  Sautéed Chicken with Tomatoes and Sweet Peppers, 64-65
  Tomato-Vegetable Soup with Boiled Chicken, 27
Chicken Liver-Mushroom Tomato Sauce, 72
Cherry Tomatoes, Sautéed, 49
Chili Sauce, English, 87
Chutney, Mixed Fruit, 92
Cioppino, 63
Clam Chowder, Manhattan, 28
Conserve, Tomato, 91
Cookies, Mincemeat Drop, 77
Creamed Tomato Soup, 24
Creole Sauce, 72

Eggplant Parmigiana, 56-57
Eggs and Cheese, Stuffed Tomatoes with, 42
Eggs in Tomato Sauce, Baked, 61

Fondue, Tomato, 57
Frappé, Tomato, 37
Freezing Tomatoes, 93

French-Fried Tomato Sandwiches, 54
Fried Red or Green Tomatoes, Neapolitan, 45
Fried Tomato Slices, 44

Gazpacho, 29
Green Tomatoes *see also* Tomatillos
  Fried Tomato Slices, 44
  Green Tomato Mincemeat, 92-93
  Kosher Dill Green Tomatoes, 88
  Neapolitan Fried Red or Green Tomatoes, 45
  Old-Fashioned Green Tomato Pickle, 89
  Piccalilli, 88-89
  Vermont Green Tomato Pie, 78
Guacamole Salad, 39
Gumbo, Seafood, 26

Ham, Eggplant, Artichoke and Tomato Sauté, 62

Jelly, Tomato, 90
Juice, Tomato *see* Beverages

Kosher Dill Green Tomatoes, 88

Lamb with Tomatoes and Spinach, 66
Lotion for the Skin, Tomato, 93
Manhattan Clam Chowder, 28
Marinara Sauce, 69
Marinated Tomatoes with Variations, 36
Marmalade, Tomato, 90
Meatballs in Tomato Sauce, 67
Mincemeat Drop Cookies, 77
Mincemeat, Green Tomato, 92-93
Mixed Fruit Chutney, 92
Mushroom Filling, Stuffed Tomatoes with, 59

Onion Soup, Tomato and, 26-27

Pastry Crust, 78-79
Piccalilli, 88-89
Pickles and Relishes
  Kosher Dill Green Tomatoes, 88
  Old-Fashioned Green Tomato Pickle, 89
  Piccalilli, 88-89
Pie, Vermont Green Tomato, 78
Pilaf, Tomato Rice, 51
Pizza, 54-55

Plum Tomatoes, Candied, 76
Pork
    Ham, Eggplant, Artichoke and Tomato Sauté, 62
    Pork with Chili Salsa, 61
    Stuffed Tomatoes with Mozzarella and Ham Salad, 41
Pot Roast, California, 66-67
Preserves *see* specific types

Quiche, Tomato, 56

Relishes *see* Pickles and Relishes
Rice Dishes
    Spanish Rice, 51
    Tomato Rice Pilaf, 51

Salad Dressings
    Tomato Salad Dressing, 32
    Vinaigrette Dressings, 33
Salads *see also* Aspics, Stuffed Tomatoes
    Armenian Salad, 38
    Guacamole Salad, 39
    Italian Salad, 40
    Marinated Tomatoes with Variations, 36
    Sliced Tomato Suggestions, 35
    Tomato Salata, 38
    Zucchini and Tomato Salad, 37
Salsa, Mexican Tomato, 75
Sandwiches, French-Fried Tomato, 54
Sandwich Suggestions, Tomato, 52-53
Sauces
    Basic Tomato Sauce, 68-69
    Canned Tomato Sauce, 84
    Chicken Liver-Mushroom Tomato Sauce, 72
    Creole Sauce, 72
    English Chili Sauce, 87
    Green Taco Sauce, 75
    Marinara Sauce, 69
    Mexican Tomato Salsa, 75
    Quick Tomato Sauce for Pasta with Variations, 70-71
    Tomato Barbecue Sauce, 74
    Tomato Orange Sauce, 73
    Tomato Sauce Provençale, 71
    Tomato-Dill Sour Cream Sauce, 73
    Western Barbecue Sauce, 74
Sautéed Cherry Tomatoes, 49
Scalloped Tomatoes, 47
Seafood
    Cioppino, 63
    Manhattan Clam Chowder, 28
    Seafood Gumbo, 26
    Shrimp Creole, 62-63
    Tomato Fans, 40
Shrimp Creole, 62-63
Sliced Tomato Suggestions, 35
Soups
    Gazpacho, 29
    Creamed Tomato Soup, 24
    Iced Tomato-Lime Cream Soup, 28-29
    Manhattan Clam Chowder, 28
    Seafood Gumbo, 26
    Tomato and Onion Soup, 26-27
    Tomato and Rice Soup, 25
    Tomato Bouillon, 25
    Tomato-Vegetable Soup with Boiled Chicken, 27
Spanish Rice, 51
Stewed Tomatoes with Variations, 50
Stuffed Tomatoes
    Stuffed Tomato Suggestions, 34
    Stuffed Tomatoes with Eggs and Cheese, 42
    Stuffed Tomatoes with Mozzarella and Ham Salad, 41
    Stuffed Tomatoes with Mushroom Filling, 59
    Stuffed Tomatoes with Turkey Salad, 41
    Tomato Fans, 40
    Tomatoes Stuffed with Tuna, 58-59
Succotash, 48

Tamale Pie, 58
Timbales, Tomato, 49
Tomatillos
    Green Taco Sauce, 75
    Marinated Tomatoes, 36
    Sliced Tomato Suggestions, 35
    Tomato Sandwich Suggestions, 52-53
Tomato Aspic *see* Aspics
Tomato Butter, 91
Tomato Catsup, 86
Tomato Conserve, 91
Tomato Fondue, 57
Tomato Frappé, 37
Tomato Jelly, 90
Tomato Juice *see* Beverages
Tomato Marmalade, 90
Tomato Paste Balls, 86
Tomato Paste, Canned, 84-85
Tomato Paste, Dried, 85
Tomato Sauce, *see* Sauces
Tuna, Tomatoes Stuffed with, 58-59
Turkey Salad, Stuffed Tomatoes, with, 41

Veal Scallops, Sorrento Style, 65
Vegetable Side Dishes
    Bacon, Tomato and Green Bean Sauté, 47
    Baked Cheddar Tomatoes, 46
    Baked Sliced Tomatoes, 45
    Broiled Tomato Halves, 46
    Fried Tomato Slices, 44
    Mixed Vegetable Sauté, Ethiopian Style, 48
    Neapolitan Fried Red or Green Tomatoes, 45
    Sautéed Cherry Tomatoes, 49
    Scalloped Tomatoes, 47
    Stewed Tomatoes with Variations, 50
    Succotash, 48
    Tomato Timbales, 49

Vinaigrette Dressings, 33

Zucchini and Tomato Salad, 37

MARGARET GIN was born in Arizona, of Chinese parents, spent most of her adult life in San Francisco and has traveled in Europe. She has translated this first-hand knowledge of the foods of three continents, combined with a remarkable talent for creative cookery, into a number of widely acclaimed 101 cookbooks: *Country Cookery of Many Lands; Innards and Other Variety Meats; Regional Cooking of China;* and *One Pot Meals.* Mrs. Gin, her husband William, and their two teenage boys divide their time between their home in San Francisco and a country house in the Napa Valley, where she teaches cooking—and raises tomatoes.

RIK OLSON, an artist versatile in many media, received his BFA degree from California College of Arts and Crafts and later spent eight years in Europe as an arts and crafts instructor for the United States Army. While he was abroad, his graphics and photographs were widely exhibited in Germany and Italy, winning a number of awards. In addition to the Edible Garden Series, Rik Olson also illustrated another book, *One Pot Meals,* for 101 Productions. He and his wife presently live in San Francisco.